Ecopedagogy

Bloomsbury Critical Education

Books in this series explore the relationship between education and power in society and offer insights into ways of confronting inequalities and social exclusions in different learning settings and in society at large. The series will comprise books wherein authors contend forthrightly with the inextricability of power/knowledge relations.

Ecopedagogy

Critical Environmental Teaching for Planetary Justice and Global Sustainable Development

Greg William Misiaszek
www.ecopedagogy.com

BLOOMSBURY ACADEMIC
LONDON • NEW YORK • OXFORD • NEW DELHI • SYDNEY

BLOOMSBURY ACADEMIC
Bloomsbury Publishing Plc
50 Bedford Square, London, WC1B 3DP, UK
1385 Broadway, New York, NY 10018, USA

BLOOMSBURY, BLOOMSBURY ACADEMIC and the Diana logo are trademarks of Bloomsbury Publishing Plc

First published in Great Britain 2021

Series design by Catherine Wood
Cover image © Studiojumpee/Shutterstock

A catalogue record for this book is available from the British Library.

Library of Congress Cataloging-in-Publication Data

Names: Misiaszek, Gregery William, 1973- author.
Title: Ecopedagogy : critical environmental teaching for planetary justice and global sustainable development / Greg William Misiaszek.
Description: London ; New York : Bloomsbury Academic, 2020. | Series: Bloomsbury critical education | Includes bibliographical references and index.
Identifiers: LCCN 2020026397 (print) | LCCN 2020026398 (ebook) | ISBN 9781350083790 (hardback) | ISBN 9781350083806 (ebook) | ISBN 9781350083813 (epub)
Subjects: LCSH: Environmental education. | Critical pedagogy. | Sustainable development.
Classification: LCC GE70 .M57 2020 (print) | LCC GE70 (ebook) | DDC 363.70071–dc23
LC record available at https://lccn.loc.gov/2020026397
LC ebook record available at https://lccn.loc.gov/2020026398

ISBN: HB: 978-1-3500-8379-0
ePDF: 978-1-3500-8380-6
eBook: 978-1-3500-8381-3

Series: Bloomsbury Critical Education

Typeset by Deanta Global Publishing Services, Chennai, India

To find out more about our authors and books visit www.bloomsbury.com and sign up for our newsletters.

I dedicate this book to my loving partner in all aspects of my life, Lauren, as well as to the rest of my family, friends, and colleagues that I cherish so very much. As this book has the goals for socio-environmental justice and peace for the entire world with planetary sustainability for all of Earth, I hope the actions that emerge from such ecopedagogical teaching will result in transformative actions towards achieving these essential goals.

Contents

List of Figures x
Acknowledgments xi
Series Editor's Preface xiii

Introduction 1
- 0.1 Key questions problematized 7
- 0.2 Pedagogical tools 9
- 0.3 Emergent from research 10
- 0.4 Structure of the book 11

Part One Introduction to Ecopedagogy

1 Ecopedagogy: An Introduction 17
- 1.1 Dialogue: Critical and authentic 28
- 1.2 Ecopedagogi<u>es</u>: Plural and intersectional 31
- 1.3 World-Earth: Complexities and conundrums 32
  - 1.3.1 Organic Ecopedagogical Modeling (OEM) 39

2 Ecopedagogical Literacy: Reading the World within Earth 44
- 2.1 Countering objectivity and neutrality: An epistemological introduction 47
  - 2.1.1 Knowledges, sciences, and technologies 52
- 2.2 Globalizations: Contested terrains 56
  - 2.2.1 Development and globalization: Introducing <u>*d*</u>evelopment and <u>*D*</u>evelopment 66
- 2.3 Economics: Countering neoliberalism 68
- 2.4 Development and citizenships: An introduction 75
  - 2.4.1 Sustainability and education for sustainable <u>*d*</u>/<u>*D*</u>evelopment (ESD) 83
  - 2.4.2 Livelihoods read with <u>*d*</u>/<u>*D*</u>evelopment 88

Part Two Foundations of Ecopedagogy

3 Freirean Reinventions: Ecopedagogy 95
3.1 Freire as an environmentalist: A very brief introduction 96
3.2 "Unfinished" world and "Finished" rest of Earth 98
3.2.1 Hope: Countering fatalism 102
3.2.2 Globalizations and fatalism 104
3.3 Countering banking education 109
3.4 Dialogue: Authentic and diverse 113
3.5 Praxis-based: Transformative 123
3.5.1 Radical, revolutionary ecopedagogical praxis 129
3.6 Bottom-up approaches 131
3.6.1 Master-slave dialectic: Consciousization of being the hosts 133
3.7 Generative themes 135

4 Teaching for Ecopedagogical Praxis: Theories, Disciplines, and Positionalities 142
4.1 Positionality problematizing 145
4.2 Radical ecopedagogical praxis through diverse disciplinary, theoretical lenses 152
4.2.1 Transdisciplinary constructions 156
4.2.2 Trans-theoretical framings 160

Part Three Teaching Ecopedagogical Reading

5 Reading through Diverse Epistemologies and Methodologies 177
5.1 Epistemologies of the South 181
5.2 Disciplinarity and research of ecopedagogical work 188
5.2.1 Research within ecopedagogical teaching spaces 196

6 Reading through Citizenships: "Development," "Livelihood," and "Sustainability" 201
6.1 Problematizing citizenships: Local-to-planetary 204
6.1.1 Economics and citizenships 207
6.2 Citizenship solidarity: Problematizing development and sustainability 211
6.2.1 Reimagining globalizations with development and citizenship 213
6.3 Citizenship solidarity: Problematizing livelihood 221

Part Four Conclusion: Ecopedagogical Possibilities and Challenges

7 Limit Situations of Ecopedagogies: Post-Truthism and the Sustainable Development Goals (SDGs) 229
7.1 Disrupting post-truthism and shoveling bullshit 229
7.2 Interrogating the Sustainable Development Goals (SDGs) 242

References 249
Index 264

Figures

1.1 World-Earth spheres 20
1.2 Photo of organic chemistry modeling kit 40

Acknowledgments

I want to thank my institution, Beijing Normal University (BNU), its Faculty of Education, and, in particular, my division, the Institute of Educational Theories, for all the support they have provided me over the last eight-plus years. This includes the funding support (the Comprehensive Discipline Construction Fund of Faculty of Education) that allowed me to finish.

It is essential to acknowledge and thank all of my colleagues and friends with whom I've had countless discussions with on ecopedagogy because the book would not have been possible without all of their critical insight. This includes, but not is limited to, Carlos Alberto Torres to an unmeasurable degree, Robert J. Tierney, Penny Jane Burke, David Yisrael HaLevi Epstein, Sondra Hale, Raymond Morrow, Massimiliano Tarozzi, and many of my colleagues at BNU, including Michael Apple, Rima Apple, Tina Besley, Michael Peters, Paul Willis, and Lorin Yochim. I would also like to thank many with whom I have had discussions on ecopedagogies, including Moacir Gadotti at the Paulo Freire Institutes in São Paulo and UCLA; Francisco Gutiérrez and Cruz Prado at their home in Costa Rica; various people at Myles Horton's Highlander Institute in Tennessee, especially Susan Williams; Douglas Kellner, Peter McLaren, and Susanna Hecht, who were on my doctoral committee at UCLA and headed by Carlos; Stephen Commins on UCLA's campus; Lesley Bartlett during CIES at Columbia University; N' Dri Assié-Lumumba in Johannesburg and NYC during WCCES and AERA conferences, respectively; Liliana Olmos and Carlos Mora-Ninci for their help in Argentina; Kai Whiting on various online calls; and Richard Kahn in Los Angeles. Of course, my ecopedagogical work would not have been possible without the environmental studies foundation that I received from my undergraduate advisor at the University of Southern California (USC), Sheldon Kamieniecki. I also want to thank all of my graduate students in my ecopedagogy, critical theories and education, and globalization and education courses, from whom I have learned so much from our in-depth discussions both inside and outside the classroom. Also, the years of support from various people at USC's Davis School of Gerontology, especially Maria Henke and Jon Pynoos, allowed me to complete my studies at USC and UCLA, as well as conduct this book's initial research.

In another category of support—and with whom I have had endless conversations on this book's topic—is my lifelong partner, Lauren Ila Misiaszek. Included also are Lauren's and my parents who I have had many conversations with on ecopedagogy—Deborah Jones, a lifelong environmental activist of Appalachia and teacher, and Charles (Chip) Jones, a Pulitzer Prize–nominated journalist who has written several books. Lastly, the support of my mother, late father, and older brother, throughout my lifetime has been unmeasurably essential in many ways, including writing this book.

Greg William Misiaszek, Ph.D.
www.ecopedagogy.com

Series Editor's Preface

This book series is being developed against an international background that comprises situations that are disturbing as well as interesting. There are concerns regarding the distribution of wealth and its concentration in the hands of a few to the detriment of the many, "the multitudes," as referred to by Michael Hart and Toni Negri.1 The series is launched at a time when the "social contract" is continuously being shredded as several people are removed from the index of human concerns. Many are led to live in a precarious state. Contract work has become the norm, a situation that renders one's life less secure. There is also criticism targeted at the very nature of production and consumption themselves with their effects on people and their relationship to other social beings and the rest of the planet.

They are also difficult times because the initial enthusiasm for the popular quest for democracy in various parts of the world has been tempered by eventual realism based on the fact that strategically entrenched forces are not removed simply by overthrowing a dictator. Far from ushering in a "spring," the uprisings in certain countries have left political vacuums—fertile terrain for religiously motivated terrorism that presents a real global security threat. This threat, though having to be controlled in many ways, not least tackling the relevant social issues at their root, presents many with a carte blanche to trample on hard-earned democratic freedoms and rights. The situation is said to further spread the "culture of militarization" that engulfs youth, about which much has been written in critical education. Terrorist attacks or aborted coups allow scope for analyses on these grounds, including analyses that draw out the implications for education.

The security issue, part of the "global war on terror," is availed of by those who seek curtailment of human beings' right to asylum seeking and who render impoverished migrants as scapegoats for the host country's economic ills. The issue of migration would be an important contemporary theme in the large domain of critical education.

We are also living in challenging times in which an attempt is made for politics to be rescued from the exclusive clutches of politicians and bankers. A more grassroots kind of politics has been constantly played out in globalized public arenas such as the squares and streets of Athens, Madrid, Istanbul (Gezi

Park), Cairo, Tunis, and New York City. A groundswell of dissent, indignation, and tenacity was manifest and projected throughout all corners of the globe, albeit, as just indicated, not always leading to developments hoped for by those involved. Yet hope springs eternal. Some of these manifestations have provided pockets for alternative social action to the mainstream, including educational action. Authors writing on critical education have found, in these pockets, seeds for a truly and genuinely democratic pedagogy that will hopefully be explored and developed, theoretically and empirically, in this series.

It is in these contexts, and partly as a response to the challenges they pose, that this new series on Critical Education has been developed. Education, though not to be attributed powers it does not have (it cannot change things on its own), surely has a role to play in this scenario; from exposing and redressing class politics to confronting the cultures of militarization, consumerism, individualism, and ethnic supremacy. The call among critical educators is for a pedagogy of social Solidarität that emphasizes the collective and communal in addition to the ecologically sustainable.

Critical educators have for years been exploring, advocating, and organizing ways of seeing, learning, and living that constitute alternatives to the main- stream. They have been striving to make their contribution to changing the situation for the better, governed by a vision or visions of systems that are socially more just. The ranks of the oppressed are swelling. Hopefully, it is the concerns of these people that are foremost in the minds and hearts of those committed to a social-justice-oriented critical education. I would be the first to admit that even a professed commitment to a critical education can degenerate into another form of radical chic or another form of academic sterility. We need to be ever so vigilant toward not only others but also ourselves, coming to terms with our own contradictions, therefore seeking, in Paulo Freire's words, to become less incoherent.

This series offers a platform for genuinely socially committed critical educators to express their ideas in a systematic manner. It seeks to offer signposts for an alternative approach to education and cultural work, constantly bearing in mind the United Nations Sustainable Development Goals that, albeit difficult to realize, serve as important points of reference when critiquing current policies in different sectors, including education. The series' focus on critical education, comprising the movement known as critical pedagogy, is intended to contribute to maintaining the steady flow of ideas that can inspire and allow for an education that eschews the "taken for granted."

This specific book in the series, by Greg William Misiaszek, highlights the ever so pertinent area of enquiry and critical education that is Ecopedagogy. It is a term which echoes the figure of Paulo Freire and the work of his followers with regard to the Carta da Terra, the Earth Charter. It focuses on a concept of citizenship that is planetary which brings to mind an earlier trail blazing book by Francisco Gutierrez and Cruz Prado (Gutierrez and Prado, 1999). In typical Freirean fashion, Greg William Misiaszek helps us read the earth and the word to stem the culture of environmental violence and degradation, rebuffing its waves and countering them with a pedagogy that underscores the interconnectedness of everything in life. The effects of corporate models of production and consumption are being felt through climate change which is making life unliveable for people located in such areas as the southern parts of Africa. Heat waves, famine, disappearance of resources and wars over access to them, the further spread of diseases such as malaria, will continue to lead to further massive waves of migration across the planet. While capital moves freely, these victims have their mobility curtailed by all sorts of legally forged manacles, lending a specific meaning to the concept of the carceral state. The urge to live, albeit frugally, makes the victims of this environmental violence risk life and limb further. Many survivors of hazards along the way (deserts, raging seas) are driven into many forms of human slavery, perhaps still clinging to that forlorn hope that keeps them striving to seek that enchanting Eldorado or that oasis of peace where their human dignity will be restored.

In the meantime, the widespread misery and degradation, the underbelly of an omnivorous global capitalism, is part of a vicious process connecting different parts of the world, where excessive environmental violence in the West is re-echoed elsewhere and falls disproportionally on people from the Global South. All animate beings are connected with any effect in one corner of the globe having its ramifications for living elsewhere. The book therefore consists of a plea to strengthen human-earth relations in the interest of sustainable development worldwide. We are all relational beings living and acting in concert with others, human and non-human within an ecological system which we have borrowed from future generations. There is so much to put right, before handing Planet Earth back to future generations, and there is little time left in which to do it. This book calls for a sense of urgency in so far as Ecopedagogy is concerned.

Peter Mayo
Series Editor,
University of Malta,
Msida, Malta

Introduction

What are the costs of environmental violence and who suffers from it? I have started many of my publications with the question I have paraphrased here, and I continue problem-posing this question throughout this book. This question highlights ecopedagogues' focus on teaching to understand all the complexities and limitations of incompleteness of answering this question fully. One of the many complexities of the question lies in understanding the "who" as all humans and populations of the world, but also widening it to the rest of Nature with the recognition that "who" might not be the best term (as will be discussed later in this section). It also lies in the recognition that the divisions of the "who" suffering from environmental violence differ, especially coinciding with sociohistorical oppressions. The use of the term "costs" is far from being one of economics—although economics is an important one; rather it refers to social injustices, planetary unsustainability, and dominance over the rest of Nature.

The environmental pedagogical model that this book focuses upon is *ecopedagogy*. Ecopedagogy is essentially literacy education for reading and rereading human acts of environmental violence with its roots in popular education, as they are reinventions of the pedagogies of the Brazilian pedagogue and philosopher Paulo Freire. Ecopedagogies are grounded in critical thinking and transformability, with the ultimate goal being to construct learning with increased social and environmental justice. Rooted in critical theories and originating from popular education models of Latin America, ecopedagogy is centered on better understanding the connections between human acts of environmental violence and social violence that cause injustices/oppressions, domination over the rest of Nature, and planetary unsustainability. Teaching to understand the social aspects of environmental issues, from local-to-global perspectives and knowledges,[1] as well as through the scholarship of multiple disciplines, is essential to determine actions for lasting changes toward environmental well-being and planetary sustainability.

As Freire taught to read and reread the world, ecopedagogues teach to read Earth with the world, as a part of Earth.[2] The term "ecopedagogical literacy(ies)" will be utilized as the critical, Freirean reading of the connections between environmental and social violence, between the world and the rest of Earth. Throughout this book the specifics and complexities of what this pedagogy and its literacy are, as well as the Freirean reinvention and his ecopedagogical work, will be unpacked. Ecopedagogy centers on "reading" environmental violence, injustices, and dominance to determine necessary transformative actions through praxis—to deepen and widen critical deconstruction of the violent acts, including through critical theorizing, to construct better environmental solutions. It will be argued that ecopedagogical work (i.e., ecopedagogical teaching, reading, and research) is necessary for deconstructing environmental actions that have harmful effects upon human societies and Earth holistically and for understanding the politics of our actions, as well as for determining sustainable actions in balance with the rest of Nature and within models of justice (e.g., social justice, economic justice, environmental justice, and labor justice).

As critical pedagogies (Gadotti, 1996; Apple et al., 2009), ecopedagogies are grounded in understanding oppressions to better determine transformative actions (i.e., praxis). What is meant by "better understanding" in ecopedagogies will be discussed throughout the book, through the aspect of deepening and widening understandings from different perspectives, ranging from the Self to local, to national, to global, to the planetary (Misiaszek, 2018b). With this widening there is the aspect of environmental well-being—of not just ourselves and our communities, but of all of human populations together and Earth overall—which, as explained by Neera M. Singh (2019), calls for an extension of NIMBY to NIABY worldwide and NOPE that has a planetary scope.

> Environmental Justice struggles are no longer limited to and framed in terms of "not-in-my-Backyard" (NIMBY) politics, rather EJ activists recognize that a logical extension of NIMBY to NIABY (not in anyone's backyard) can only be achieved through a loud and clear "NOPE" (not on planet earth)—a clear refusal to activities that generate environmental harms and demand sacrifice zones.

The question facing environmental pedagogical work is this: How can we teach, read, and research through more local, contextual NIMBY, and widen it, still contextually, to NIABY and NOPE?

Throughout this book I will return to Singh's (2019) argument within the realm of economics and connections to constructs of what is "development," in that necessary environmentalism can only be accomplished through "radical restructuring of our economies and rethinking our obsession with growth."

A crucial question in teaching for "progress" is the question of whether the development and its processes are bettering the current situation. If "yes," for whom—with the expansion of "who" as all of Earth, as discussed previously?

In this book I will describe critical, Freirean-based ecopedagogies for teaching the often hidden politics between environmentally negative actions and social violence to deepen and widen (to global and planetary perspectives) students' and teachers' socio-environmental literacies. This includes "reading" what development is in order to help lead to transformative actions for planetary justice and for globally inclusive development.

Examples will be given throughout this book to explain how ecopedagogues can teach for deeper local-global understandings for praxis in framing development to emerge for globally inclusive socio-environmental justice and planetary sustainability. Education for Sustainable Development (ESD) largely emerged from Environmental Education (EE) models to focus teaching on how actions in the name of "development" negatively affect our environment and, in turn, negatively affect societies (socio-environmental), so that students understand that development should not cause socio-environmental oppressions in the present or future. However, ESD models have been increasingly taken over by the "development" framing of ESD within neoliberal economic models, rather than within economic justice models (Misiaszek, 2020, 2018b, 2011). Development is often taught non-contextually, singularly framed as Western constructs of development largely intensified by globalization from above. One key focus in this book will be how globalization's effects upon socio-environmental issues must be taught contextually at local levels (i.e., deepened understandings). Ecopedagogical work has a specific focus on teaching how *othering* intensifies socio-environmental oppressions by using various critical theoretical lenses, such as ecoracism, ecofeminism, and (neo)coloniality. With intensifying globalization, these power structures are global in scope and affect societies locally, calling for the use of global lenses (i.e., widened understandings) and local lenses to understand socio-environmental violence.

This book provides a critical analysis of reinventing environmental pedagogies for transformation through inclusivity of all the diverse societies of the world and balance in the rest of Nature, beyond humans. Critical problem-posing of the deeper aspects of and rationales for developmental goals from various perspectives, rather than a singular universal one, is essential in ecopedagogical work. Ecopedagogical work calls for deconstructing "development" in order to reconstruct "development" to be grounded in social justice, non-domination of the rest of Nature, and local-to-planetary sustainability. This calls for reinventing not only how we define development but also how we define sustainability, as I have expressed in a previous work:

> "Development" is often not critically debated contextually but is frequently touted as the reason for negative environmental actions, but are the actions truly "development" if environmentally unsustainable? "Sustainability" is often brought into the argument to limit such actions but it is often overshadowed by economic development, especially within neoliberal development models in which processes of globalization from above is helping to normalized worldwide. . . . Together, "sustainable development" is very important concept to understand to guide action but only when it is taught critically, within a planetary framing, and is contextual through socio-environmental from local-to-global to planetary understandings. (Misiaszek, 2018b)

As ideologies of "development" and "progress" largely determine our actions, how do we frame "development" with the corresponding term "sustainable development"? Schooling is often viewed as having the goals of "development" and "citizenship," with the latter having many terms that link back to how John Dewey (1963) described that "community or social group sustains itself through continuous self-renewal, and that this renewal takes place by means of the educational growth of the immature members of the group." Whose "development" factors into others' de-development, within both similar and differing framings of what the development goals of different populations are (Misiaszek and Torres, 2019; Misiaszek, 2020, 2018b, 2020a,b)?

Citizenships and citizenships education will also be a key focus of this book's discussions, with the pluralization of citizenship<u>s</u> indicating differing spheres of citizenship from local-to-global, especially with intensifying globalization. Included in these discussions is the problematizing of "development" and "sustainable development" with citizenship, especially sociohistorical oppressions, including environmental injustices, placed upon "noncitizens." A theme that emerged from research I have conducted was how environmental rights and responsibilities are elements of individuals' citizenship, with the argument of citizenship and environmental pedagogies being inherently dependent (2018b, 2016a, 2015). Globalization widens an individual's citizenship with multiple and often conflicting citizenship spheres (e.g., local, national, global, and/or planetary citizenship spheres), which further complicates the localization of environmental rights and responsibilities. I will discuss many of these complexities with specific focus on global citizenship/education (GC/E) and planetary citizenship/education (PC/E). In the same way in which processes of globalization form a contested terrain of possibilities for empowerment or oppression through education, (dis)connections between environmental pedagogies and GCE models will be problematized throughout the book. This problematizing is currently an important topic as UNESCO and other international entities have focused on GC/E. In previous publications, I have argued through other scholars' work that all schooling is for two goals with citizenship as one of them—although the term itself might not be invoked specially, schooling to before parts of society is essential. Unfortunately, citizenship has historically also been utilized as a tool for oppressing those deemed lesser or noncitizens, a topic that will be discussed throughout this book in relation to ecopedagogical work, especially colonialization and othering.

Returning to the arguments of "development," schooling overall had the goal of teaching for development but ecopedagogical work centers the question of *whose* "development" the actions in the name of development are for. This problematizing also extends to the question of sustainability, in problem-posing the question of who and what are focused upon when we invoke the term "sustainability." In addition, ecopedagogical teaching, as well as research, must be problematized within larger planetary understandings of how human actions affect the rest of Earth beyond, although very much including, anthropocentric lenses (planetarity). In the subtitle of the book the term "Global Sustainable Development" reflects all-human inclusivity of development and this is within "Planetary Justice" models including social justice (i.e., humans, the anthropocentric sphere) and all the rest of Nature (planetary sphere).

Now I will return to deconstructing the key question that begins this introduction. First is the "who" in the phrase "who suffers from it?" Coinciding with critical and postmodern theories that help to better understand and teach what populations are oppressed within societies and educational structures as inseparable parts within and between societies, ecopedagogues use these theoretical lenses to understand the causes and effects of oppression due to human actions that cause environmental violence. There are two key aspects to know before briefly describing the use of theoretical lenses. First is that when the term "education" is used here and throughout the book, the discussions are within the broader aspects of how we, as humans within and between populations, "learn" anything with formal, nonformal, and informal learning spaces.

Second is that ecopedagogies, although plural in definition/framing, have a similar grounding focus on human actions that cause environmental violence. The word "cause" is used in the previous sentence, to signify how environmental violence affects populations, but the violence itself exists due to human actions. In other words, by utilizing an example, a "naturally" occurring hurricane would not fall into this category; however, this would assume that the hurricane, its intensity, and/or the frequency of hurricanes' occurrences were not altered by humans' actions. There are many causes of and additional factors for hurricanes, but one that is likely to come to most peoples' minds is the effects of human-caused climate change. In addition, this book aims to problematize how we have constructed our environments, with all the complex power relationships of those constructions, in the name of "development" and "progress," which oppress some populations but not others.

This book makes use of theoretical lenses to better understand "who" is most socio-environmentally suffering (in the question beginning this book's introduction), with such suffering largely aligning with sociohistorical oppressions within the anthropocentric sphere. Such oppressions coincide with power structures within and between societal divisions, including, but not all-inclusive of, class, gender, race, sexual orientation, ethnicity, nationality, and geopolitical location. All these socially constructed divisions, as well as others, will be discussed through this book as points of socio-environmental oppressions but also possibilities of empowerment and environmental well-being.

The wording that ends this last sentence gives way to the next complexity in defining the "who" in the question. For the sake of simplifying the question's wording, while acknowledging the linguistical limitations of the word "who," I expand the who beyond the anthropocentric sphere of humans. The "who" in this case could be popularly worded as "what"; however, the "what" also objectifies all entities outside of humans. This coincides with the wording of "Earth" without the "the," as briefly described in the second endnote. With imperfect limitations of the English language to describe Earth's entities, the "who" remains with the acknowledged aspect of being problematic, for example, the issue of humanizing all the rest of Nature (humanizing nonhuman animals with "human" characteristics).

Earth is heading toward multiple environmental crises that are more devastating than the sum of separate crises; however, environmental pedagogies have not radically transformed the way most people view our relationship with the rest of Nature, or with each other as humans, to drastically alter actions needed to save the planet. For example, most people understand the threat of global warming but actions to curb it often conflict with false-normalized neoliberal framings of "development" (or "progress"). These normative framings often counter what is essential for environmental-social transformation; however, we continue to guide our actions by such framing of "development," such as neoliberal ones, even though it is not true development/progress. Education constructs how we view the importance of a healthy environment for development and thus guides our actions affecting Nature. Education here is used in the widest respect, including formal, nonformal, and informal learning spaces, of how individuals learn to understand anything and everything. Without teaching the inherent connections between environmental and social violence (socio-environmental violence), development is too often separated from environmental well-being and decision-making based on development framing that separates us (i.e., humans, human populations) from the rest of Nature. This book discusses how environmental teaching must focus on connections between diverse societies with all that is Earth to critically determine what sustainable development is and how to construct possible ecopedagogies.

0.1 Key questions problematized

Returning to the previous example of global warming, ecopedagogues problem-pose how we are taught to continue to engage in environmentally violent actions, especially at the current pace. This question does not arise from ignorance, but from the current knowledge that we know we will ultimately ruin our home—Earth. What current teaching and element absent from teaching allows this? In answering this question on how education must change, the following are some overall questions that will be examined in this book:

- How can we teach environmental issues that can help lead to actions to save all of Earth?
    - Why are current environmental pedagogies often failing at this?
- How do educators, in general, and environmental pedagogues teach to understand or ignore, misunderstand or misteach consequences from environmental violence?
- How is the rest of Nature taught to be "understood" in relation to us as humans?
    - How do understandings lead to environmental peace, violence, and/or un/sustainability?
- What are the politics for teaching to mislead people on environmental violence causes and effects?

- How can we teach to recognize such misleading teachings, and what are their reasons?

- What populations are most affected (e.g., gender, race, culture, global location, migration status, (non)nationality, ethnicity, sexual orientation, (non)religiosity/spirituality) by environmental violence and why?
  - How are these factors taught or not taught? And why?
- How does the intensifying processes of globalization (oppressive and empowering) affect environmental teaching?
  - How are histories of oppression, especially histories of colonialism, taught or not taught as continuing factors of socio-environmental oppressions from globalizations (e.g., globalization as neocolonialism)?
  - How can we teach to disrupt socio-environmental oppressive processes of globalization and utilize globalizations for world-Earth peace holistically?
- How is "development" and "progress" taught, and how do we frame who benefits from such "development" and "progress"?
  - How is "sustainability" taught within the realm of development and the importance, or not, given to the rest of Earth?
- What knowledges and epistemological framings (i.e., ways of knowing) do we use or not use to understand the diverse world and the rest of Earth?
- How do taught framings of "citizenship" both connect "us" to one another and separate "us" from "them" (i.e., the "noncitizen")?
  - How do we teach, or not teach, widened and narrowed spheres of citizenship (from local to national to global to planetary citizenships) as caring for one another and the rest of Nature, or as not caring?

In answering the problem-posing questions above through ecopedagogical work, we as learners and teachers help to determine the politics of actions that lead to global warming, as well as other actions that lead to unsustainable environmental violence. False *hidden curricula* environmental violent acts result in maldevelopment and local-to-planetary unsustainability, teaching an ideology in which such violent acts are seen as "necessary evils." This, in turn, only aids in sustaining and often enhancing environmentally violent acts by minimizing or ignoring socio-environmental negative effects in teaching. It will be argued that teaching for deepened and widened understandings of who are negatively affected and who benefits from environmentally ill acts is essential in how we teach for true, planetary holistic development and, thus, actual sustainable development.

0.2 Pedagogical tools

Answering the overall question at the outset of this introductory chapter is inherently complex, although the question's wordings are simple. Environmental pedagogies that lead students in answering this question with simple, mostly binary answer(s) are inherently shallow and noncritical. Such teaching is non-transformational and helps to separate environmental violence from social violence, and often from planetary sustainability—opposite to ecopedagogies. Effective ecopedagogical models are essential for understanding and constructing solutions for socio-environmental injustices because false environmental knowledges are so deeply hidden by prevalent noncritical environmental pedagogical models. Such noncritical models end up, at best, intensifying injustices by ignoring socio-environmental connections and, at worst, solidifying ideologies that the oppressed are benefiting from what is oppressing them the most.

Building on these arguments throughout this book, I will discuss how, and the reasons why, many environmental pedagogues fail to teach in ways that result in students taking substantial actions to end unsustainable environmental violence. In this book I argue that environmental pedagogies must be critical in essence, with democratic, dialogical teaching through problem-posing, often inconvenient, and sometimes uncomfortable, questions that interrogate students' and teachers' ideologies. This book delves into how structural aspects of schooling and education form politics that directly and indirectly oppose ecopedagogical models. Critical environmental pedagogies are often avoided due to larger systematic politics rather than by teachers themselves. Noncritical environmental pedagogical models are not only ineffective but also help to sustain socio-environmental oppressions by avoiding teaching that questions power entities and inequalities (e.g., hegemony) that are frequently the sources of environmental violence.

I will provide various pedagogical tools, research agendas, analyses, and core foundations on how to ecopedagogically reinvent environmental teaching to be more effective in achieving global social justice and planetary environmental justice. These discussions are not meant to give overall answers or solutions, especially in step-by-step ways, but will provide the foundations for successful ecopedagogy that needs to be locally contextualized. In other words, ecopedagogies are not singular or universal, but contextualized within societies with core tenets, rather than being static, technocratic implementation. This contextually is foundational in ecopedagogy.

Many current environmental pedagogies oversimplify environmental issues by placing them into single disciplines, through a single epistemological lens, and through a single theoretical lens (if any, at all) (2018b, 2011). As my previous research has shown, theoretical, epistemological, and disciplinary framings that environmental issues are taught, researched, and read through are often too simplified leading to narrowed and shallow learning (2018b, 2016a, 2015, 2014c, 2012, 2011). Ecopedagogies must be transdisciplinary, trans-epistemological, critically comparative, problem-posing, and trans-theoretical toward constructing possible environmental solutions through democratic, dialectical teaching, to name a few key tenets. Through theoretical lenses and by constructing theoretical frameworks I will discuss how both students and teachers can begin to understand complex socio-environmental issues that unequally oppress different populations toward praxis (constructing effective action through theoretical reflection). Some examples of theoretical lenses that will be discussed in the book include theories of globalizations, critical race theories, theories of feminism, citizenship, coloniality, and media culture theories.

0.3 Emergent from research

Although not meant to be a book that directly documents research, this book will be grounded, in part, in two comparative/international ecopedagogical research projects. The first one was conducted in Brazil, Argentina, and the Appalachia region of the United States, which all have long histories of popular education due to the influence of such educators as Freire (O'Cadiz et al., 1998; Au, 2009) and Myles Horton (Horton et al., 1998, 1990; Adams and Horton, 1975). Through qualitative research with several expert ecopedagogues, this book will offer what the Global North can learn from the Global South, and vice versa, on ecopedagogical work (Misiaszek, 2012, 2011).

A second research project on critical scholars working in various aspects of globalization and education with internationally recognized expertise in citizenship pedagogies and/or environmental pedagogies will also be utilized indirectly throughout the writing (2016b, 2015, 2014b). The participants, working in six continents, were asked open-ended questions focused on the following four topics: framing citizenships and pedagogical aspects; essential aspects of environmental pedagogies; connections between citizenship and environmental pedagogies; globalization's effect on these connections and disconnections; and what the essential pedagogical changes in/between both fields are. These research analyses will indirectly offer various possible solutions that emerged from the participant voices that discussed, for example, effective ecopedagogical tools; necessary systematic changes in environmental pedagogies at all levels (formal, informal, and informal); essential international environmental educational policies reform; reinventions of local, nation-state, global, and planetary citizenship; and centering deep theoretical analysis in teaching to provide multiple lenses and diverse perspectives of socio-environmental issues.

Although not a focus and not discussed at length, this section notes that this book's contents emerged from rigorous qualitative and theoretical research.

0.4 Structure of the book

The linear format of writing, such as in this book, makes writing about the nonlinear, complex connectivity between the world and Earth extremely difficult and challenging. This includes, but is also far beyond, writing on ecopedagogical work; however, this is the task at hand. I state this complexity and problematize the linearity of writing to tell the readers that there will be referencing back and forth of the different ecopedagogical concepts—ones that have already been described or will be described later—as well as instances when some concepts will not be fully explained within this book's pages owing to limitations of space. For this latter aspect of incompleteness, I have tried to offer extensive citations, references, and footnotes throughout the book for guidance on additional information and explanations. With this nonlinear*ness* discussed, I have structured this book in one of the many ways this book could have been written.

The first part of the book ("Introduction to Ecopedagogy") will introduce the overall key aspects of ecopedagogy, including those of ecopedagogical literacy. This first part will begin the discussion on the need to problematize "development" and "sustainability" in an increasingly globalized world. I will critically deconstruct the need for contextual environmental pedagogies to dialectically understand the positive and negative effects of globalization<u>s</u> ("from above" or "from below") with the complexities of the needs of the rest of Earth (beyond anthropocentric justice models), to construct effective ecopedagogies in an increasingly globalized world. Included in this part is a discussion on how ecopedagogies and ecopedagogical tools can be used with, as well as within, other environmental pedagogies such as EE and ESD models.

Questioning "development" is done by problematizing who is positively or negatively affected by framings of development, and the often conflicting needs of Earth holistically. It includes the problem-posing of what is "sustainability" and what local-to-planetary sphere(s) we must consider when determining sustainability in our actions for development. In addition, there are the complexities that stem from the fact that humans develop through their self-reflective actions while the rest of Nature adapts, positively and/or negatively, without the ability to reflect. This will be analyzed throughout the book, with the following quote by Freire on the uniqueness of human abilities for sociohistorical analysis to transform the world, as we make up the "world."

> I shall start by reaffirming that humankind, as beings of the praxis, differ from animals, which are beings of pure activity. Animals do not consider the world; they are immersed in it. In contrast, human beings emerge from the world, objectify it, and in so doing can understand it and transform it with their labor. (Freire, 2000)

With development as defining the goal(s) and pathways of transformation, a key question is "who" is included in benefiting from "development" and what ideologies and theories are or are not reflected upon in its constructions (i.e., praxis).

In his later writings, which will be discussed throughout this book, Freire wrote on the need for transformation through praxis encompassing the rest of Earth. This part will discuss the connections and disconnections between local, global, and planetary lenses for ecopedagogical teaching and research. A key aspect that will be explained and theorized at length is the defining of the world (i.e., all humans, the anthropocentric sphere) and the non-anthropocentric sphere (i.e., all of Nature outside of humans) as making up Earth (i.e., the planetary sphere, all of Nature) holistically.

Part II, "Foundations of Ecopedagogy," will delve into the Freirean foundation of ecopedagogies and teaching for praxis with disciplinary, epistemological, and theoretical aspects introduced. The essence of ecopedagogical work as radical will be emphasized throughout this part. This second part will include discussions on how critical theories are necessary through ecopedagogical curriculum, teaching practice, research on environmental, and/or sustainable development education.

The third part ("Teaching Ecopedagogical Reading") returns to the topic of ecopedagogical literacy to further deconstruct the pedagogies, epistemologies and methodologies we utilize to read our environmental violence, including what is "citizenship," "development," "sustainability," and "livelihood." In some of my previous publications, I have written on necessary connections between ecopedagogy and citizenships education (multiple spheres of citizenship from local-to-planetary is denoted as the plural form citizenships throughout this book). Spheres of citizenship will be a central type of analysis in positing who we, as humans, develop with/in sphere(s) including traditional national citizenship, subnational citizenship (to local communities), to supranational including aspects of global citizenship and its education (GC/E) and planetary citizenship and its education (PC/E).

The fourth and last part ("Conclusion: Ecopedagogical Possibilities and Challenges") concludes by pulling together some of the key ecopedagogical aspects within the current era of post-truthism, along with some very brief thoughts on the UN Sustainable Development Goals (SDGs). Throughout this book there is the overall focus on how environmental pedagogies (e.g., EE, ESD) need a paradigm shift in order to teach to critically deepen and widen understandings of the socio-environmental connections for transformative actions to save Earth. This also includes the ecopedagogical analysis needed of various environmental pedagogical initiatives, such as the SDGs.

There are many aspects of ecopedagogical work, as discussed in this book, that are essential to transform environmental pedagogies, including how we teach, read, and research on the SDGs. This includes the deconstruction and possible reinventions of the SDGs, as well as the connections between the 17 Goals, as both these goals and their official wording coincide and contrast with ecopedagogical tenets. Although brought up throughout the text, there is no extensive direct analysis of the SDGs specifically in this book, but all of the ecopedagogical work serves as tools and foundational tenets that ground problematizing of SDGs—both their possibilities and challenges. For those who read this book for, in part, UN SDG analysis, the following three overlaying questions should be problematized: How is "development" defined in the SDGs? What is emphasized and why? And what is missing in the SDGs overall and with the SDGs individually, and why?

Notes

1 The plural form of knowledges signifies that individuals have multiple, and often contested, understandings that are culture based (e.g., local knowledges, Indigenous knowledges) and those emergent from dominant ideologies.

2 Throughout the book, the term "the earth" will not be used to de-objectify Earth with the article "the" or the lower case "e." Aligning to this argument, Nature will be uppercased throughout this book.

Part One

Introduction to Ecopedagogy

1

Ecopedagogy

An Introduction

To stop the downward spiral of intensifying environmental violence that leads to social violence, we need to better understand what the politics of humans' violent actions are in order to determine why the violence is taking place (i.e., the root causes). The overall question problematized throughout this book is this: How can environmental pedagogies end socio-environmental oppressions and planetary unsustainability? What pedagogies are necessary for "true" sustainability for all of human societies (i.e., the world) and all else that is Earth. The root causes of environmental violence are most often purposely hidden, to sustain and intensify the benefits of hegemony for a few. Ecopedagogies have the overall goal of unveiling these hidden politics—for teaching, reading, and research toward this end.

This chapter will discuss the overall key tenets of ecopedagogy, as well as introduce some foundational Freirean aspects of ecopedagogies, with discussion on how environmental pedagogies (dis)connect us, as humans making up the world, with the rest of Nature, making up Earth. Beyond "environmental pedagogies," ecopedagogical work also includes critically deconstructing *pedagogies on the environment*, with this phrase indicating education which can either be for environmental "well-being" or cause environmental destruction. Stated a bit differently, "pedagogies on the environment" will denote any type of teaching on the environment, but this does not necessarily mean that the underlying foundation is environmentalism, either intentional or not, or consciously or not. This is different from *environmental pedagogies*, in which environmental well-being is the goal, regardless of whether or not the teaching is effective. Ecopedagogical work deconstructs if pedagogies on the environment models are environmental pedagogies, in which what is taught on the sustainability between the world and Earth is truly toward this goal. Ecopedagogical work deconstructs the politics of the former and the latter, to see who truly benefits and who is negatively affected within the ideologies of the *hidden curricula* (e.g., what is indirectly being taught through curricula), as defined by Henry Giroux (2001) and others. The discussions of these topics will begin here and will continue throughout this book, including the next chapter on teaching literacy to read the world as part of Earth (i.e., *ecopedagogical literacies*) and Chapter 3, which will focus on the reinvention of Freirean scholarship for ecopedagogical teaching and scholarship to emerge.

As ecopedagogy is reinvented from Freire, it is important to remember that Paulo Freire was a teacher of literacy—teaching to read the word in order to read and reread the world. Reading is not passive, to only gain understanding, but to critically determine action. As Freire's (1997) quote indicates, reading is for transformational praxis through better understanding the world, rather than to know the world as non-transformable (or static and fatalistic with a predetermined future).

> One of the fundamental differences between me and such fatalistic intellectuals—sociologists, economists, philosophers, or educators, it does not matter—lies in my never accepting, yesterday or today, that educational practice should be restricted to a "reading of the word," a "reading of text," but rather believing that it should also include a "reading of context," a "reading of the world."

Critical/Freirean pedagogies are not only to understand socio-environmental issues but for praxis. Praxis is a complex and multifaceted goal as Moacir Gadotti (1996), among others, has argued, which will be further discussed throughout this book, but here, in the field of ecopedagogy, it is how to teach for learners'/readers' transformative actions through deepened and widened reflections toward ending socio-environmental injustices, violence, and dominance over Nature.

As I write "Earth" without the article "the," as in "the Earth," it is not in error but rather so as to not objectify Earth as an object, as within tenets of ecolinguistics (Stibbe, 2014, 2012; Derni, 2008; Fill, 2001). The "world" signifies all humans and their groupings, which will be discussed further in this chapter, but a key aspect of ecopedagogy and its literacy is deconstructing social and environmental violence in the world *within* Earth. Teaching to understand, or crucially read, the connections between social violence/injustice and environmental violence, which are inherently inseparable, is the goal of ecopedagogies (Gadotti, 2008b, 2008c). These connections are often structurally hidden to conceal who really benefits and the many more negatively affected. This is with the assumption, which I have introduced elsewhere, that all environmental violence happens to benefit somebody, with ecopedagogical teaching to read what the politics of the degrees of who benefits and who suffers from environmental violence are (Misiaszek, 2018b). This assumption is also from the work of Michael Apple.[1] Ecopedagogues teach to read the politics of these connections, and ecopedagogical research seeks the "hows" and "whys" of these connections not being taught or being mistaught in education (formal, nonformal, and informal education; environmental pedagogies and pedagogies on the environment) (Misiaszek, 2018b, 2011). A topic that will be discussed more in length later is that ecopedagogical research is inherently being done in ecopedagogical learning and reading spaces. Throughout this book I will utilize the term "ecopedagogical work" to indicate ecopedagogical teaching, literacy/reading, and research—only naming the specific type when directly referenced among the three.

Ecopedagogical work attempts to deconstruct the continuum of environmental violence to devise transformative actions that can be undertaken by students to end environmental violence and achieve sustainability at all levels from local-to-planetary. What I mean by a continuum of environmental violence is that we, as the world, utilize the rest of Earth for our own needs and wants. If there was no benefit for anyone, humans would not cause environmental violence. For example, why would we drill deep into the oceans' floors for a slippery, viscous fluid (i.e., oil) unless it served some sort of purpose? Oil serves many purposes such as for lubrication and as the fuel we use every day. Of course, there are many, many factors of oil such as unequal usage, distribution, and almost endless pollution effect, but I will get to those later in the book. This example illustrates a key factor of ecopedagogical work for better reading to unveil the politics of oil mining and consumption that lead to environmental violence and, in turn, lead to social violence. There are almost limitless questions to problematize within this example. One is questioning what is "needed" compared to what is "wanted," especially within social justice models. Another is what the other possible sources of energy are and the politics behind oil remaining a primary source, with the accompanying question of who benefits from this (e.g., deconstructing the politics of oil production).

There are also questions about the effects of our environmental violence beyond humans. Ecopedagogical work also examines things that are beyond the world (i.e., beyond humans, beyond the anthropocentric sphere)—all else that make up Earth. In this book, the non-anthropocentric sphere will indicate all else that is Earth beyond humans. Other terminologies that will be utilized are within the concepts of "spheres," with Earth as the planetary sphere that includes all humans of the anthropocentric sphere (i.e., the world) and the "non-anthropocentric sphere" as the planetary sphere not including humans (i.e., the anthropocentric sphere). Ecopedagogical content and goals are infinitely complex within and between these spheres, among which this book will address some, but all-inclusiveness is impossible. To help guide some of the key aspects of ecopedagogical work, Figure 1.1 illustrates the vast scope and the overall goals of such work.

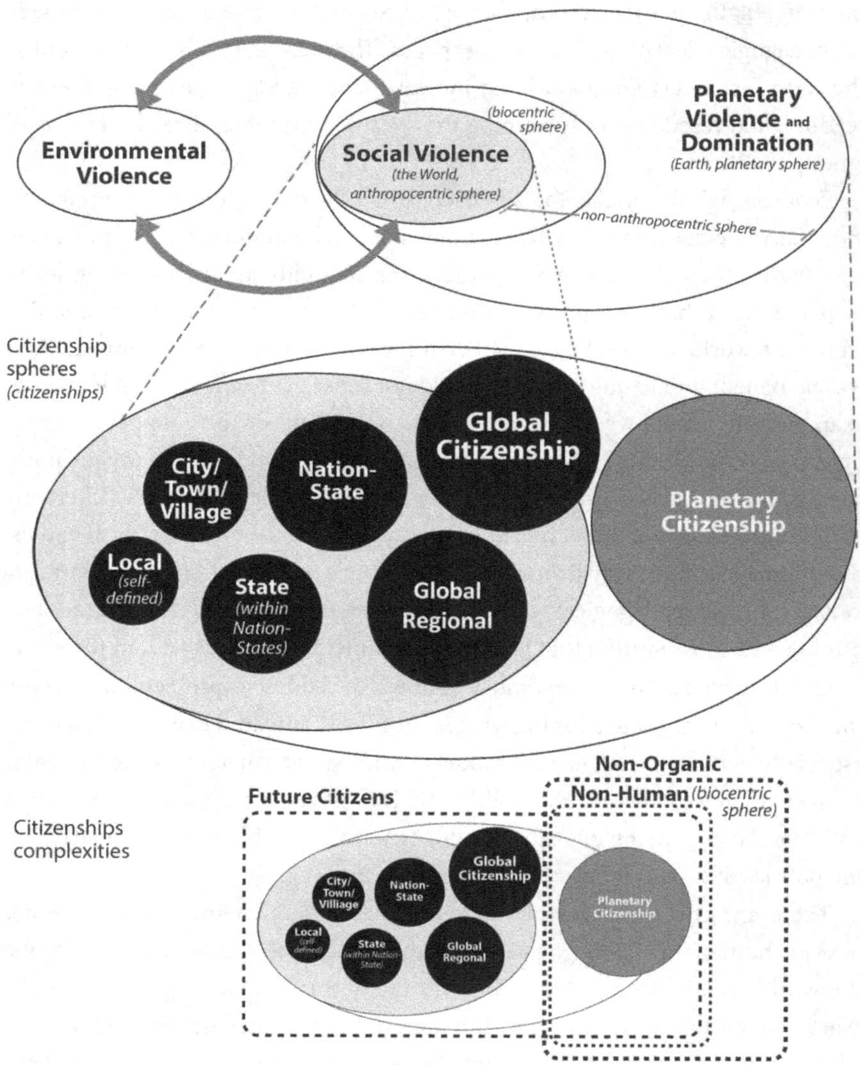

Figure 1.1 World-Earth spheres.

The three interrelated graphics in Figure 1.1 have evolved through my ecopedagogical work, with other versions of it given in previous publications (2018b, 2011). Ecopedagogical work focuses on the connections between environmental violence and social violence, as well as how environmental violence affects everything else that makes up the planetary sphere, as illustrated in the figure. The imagery of the connections between environmental and social violence have figured throughout my work; however, the graphic illustrates the world within Earth with some key aspects of the complexities of the ways in which effects of development and sustainability must be widened to the planetary sphere (i.e., Earth holistically). The planetary sphere is illustrated as divided between all that is living with the label "biological sphere" that includes the world (anthropocentric sphere) and the rest of Earth. Throughout this book, the sphere of Earth outside of the world will be called the "non-anthropocentric sphere." However, these divisions as "hard" boundaries are false, as between the world (anthropocentric sphere), the biological sphere, the non-anthropocentric sphere, and the rest of Earth. These spheres are intertwined together with limitless complexities between us, as humans, within the larger biological sphere, and within the complexities of the planetary sphere. Figure 1.1 will be referenced throughout this book, especially in the discussions on citizenships, which are denoted by the bottom two illustrations in the figure.

The graphic illustrates also the complexities of "who" is included within the question of "who" benefits and "who" suffers—for example, pondering if the inclusion of "who" is within the world or rather also beyond the world to include Earth holistically.[2] Where in graphic is/are the "who" represented, as an entity(ies) that is oppressed/dominated by environmental violence, are educational models concerned with? As stated previously, critical pedagogies have the goal of understanding oppressions better, with specific foci on who suffers the "most," but ecopedagogies expand this determination beyond humans to the planetary sphere. However, as described in the next section, the essence of ecopedagogical work is to understand oppressions from those who suffer the "most" and are dominated within the non-anthropocentric sphere.

Ecopedagogical work coincides with critical pedagogical work (even more so specifically with Freirean pedagogies, which will be discussed throughout this book), with the focus on teaching about human actions that are acts of environmental violence. Environmental violence emerging from the world, as human actions, is an essential defining factor of ecopedagogical work in that "naturally" occurring factors, such as hurricanes, forest fires from lightning, and earthquakes are not focused upon. However, the reason for emphasizing "naturally" in the previous sentence is because those environmental disasters can be caused by humans (e.g., storms intensifying as a consequence of global warming), and/or the disastrous effects might be mostly within the anthropocentric sphere (e.g., human-made structures affected by fires and earthquakes), making ecopedagogical work much more complicated. So ecopedagogical work problematizes why we, as humans, commit acts of environmental violence.

This book will largely focus on environmental violence done in the name of "development" and "citizenship"—both being reasons for promoting and/or lessening environmental violence toward planetary sustainability. It is human actions that directly or indirectly lead to environmental violence as we, as self-reflective, historical beings, can dream of utopias (three characteristics that Freire (2000) expressed as unique to humans). The (in)direct aspects are complex and make the "hiding" of the true reasons for social-environmental violence unfortunately easier. This is one of many ways, among others that will be discussed throughout this book, in which "hiding" socio-environmental connections results in less protest against environmental violence, calling for urgent ecopedagogical work.

"Development" and "citizenship" are often seen as primary reasons for public education to emerge; however, most critical pedagogues argue that the ultimate aim of schooling historically has often been social control and sustaining power stratifications (especially hegemony) rather than all-inclusive well-being, democratic participation, and empowerment. Problematizing constructs of development is essential within ecopedagogical work to question whose development it is and what its effects, ranging from the local to the planetary, are. As will be discussed later, the importance of development can be witnessed as the "D" within the environmental pedagogy ESD, but what ecopedagogical work focuses on is problematizing what the "D" is, especially in relation to the "S" (i.e., Sustainability). Is development being taught as local-to-global all-inclusive progress, or is it progress for a few and, if yes, why? Beyond this question is how such a "development," or de-development, coincides with planetary sustainability or unsustainability. Questions of (un)sustainability are ones that must be problematized within the planetary sphere toward Earth's holistic balance. Returning to the notion that environmental violence is a continuum resulting in both benefits and oppressions to the world, ecopedagogical work on planetary sustainability problematizes what the balance of environmental violence from local-to-planetary levels is (Misiaszek, 2018b). As will be discussed in the book, ecopedagogical work deconstructs how different populations are socio-environmentally affected differently, often coinciding with sociohistorical oppression divisions (e.g., gender, race, ethnicity, sexual orientation, non-/nationality(ies), and migration status).

Throughout this book, I will argue, through previous research that I have conducted, that often environmental pedagogies do not teach through such problem-posing of environmental violence (Misiaszek, 2020, 2018b, 2011). I will unpack how education, and often environmental pedagogies, often widens the gaps between our "development" and the rest of Nature, as we increasingly politicize the "laws of Nature" even though they are static and unadaptable to the world's politics. "Development" is often taught outside of, or challenging, rather than within, Nature's laws. It is important to note our limitations in fully understanding Nature's laws, but here I am arguing against the blatant unsustainable, de-development actions ignoring Nature's law's that we know, to the highest degree of certainty, are true (e.g., climate-change deniers).

In relation to the other educational connection to citizenship, I will continue to argue the need to both deepen and widen citizenship from the local-to-planetary levels. Throughout this book I will discuss the need for ecopedagogy within and between multiple spheres of citizenship (denoted as the plural form citizenships, see Figure 1.1 middle and bottom illustrations, citizenship spheres (citizenships) and citizenships complexities), with particular attention to "planetary citizenship" in solidarity with the rest of Earth, with the goal of sustainable development. Within citizenships it is important to have solidarity with one another as we are all connected with Earth holistically, acknowledging the complexities of respecting not only the vast diversity between citizenships but also the commonalities between them. When I discuss global and planetary citizenship it is not to devalue the contextuality of other citizenship spheres, including traditional national citizenship, but rather to emphasize the interconnectivity between them, especially with intensifying globalizations—each other's actions, no matter the distances between them, affect each other and Earth overall. To counter "development" within problematic citizenship framings in which development for "our society" is on the backs of "their society," critical global citizenship models and planetary citizenship models widen the "our"/"us" to mean all humans and "their"/"them" to be inclusive of all of Earth.

The Brazilian scholar Moacir Gadotti (2008a, 2008b, 2008c), a former doctoral student of Paulo Freire, defined planetary citizenship as "an expression that was adopted to express a group of principles, values, attitudes and habits that reveal a new perception of Earth as a single community". Nature within the non-anthropocentric sphere has the intrinsic "resting state" of being balanced, with only humans as agents of imbalance, as we are self-reflective organisms that consciously act against returning to equilibrium, challenging Earth's balance (Misiaszek, 2020, 2018b). Although we are the only beings on Earth that can understand and reflect upon the negative consequences of such unbalance/unsustainability, we can also choose to ignore such consequences. Ecopedagogical work is to counter such ignorance, with the specific concern of problematizing how societal structures of the world often systematically suppress knowledges and critical thinking for ignorance that we are not conscious of such systematic intentions. Educational systems (formal, informal, and nonformal (e.g., public pedagogies)) are such key structures in sustaining such ignorance and/or critical consciousness of socio-environmental and unsustainable actions. Ecopedagogies are inherent enemies of educational systems that politically teach socio-environmental ignorance, as the former inherently counters such systems. The (in)direct ways of teaching such ignorance, as well as deconstructing the political reasons behind them, will be a key topic throughout this book.

Deconstructing the "ways" mentioned in the previous sentence is an endless task. For example, how do our ways of knowing come about through education, with the realization that the world's diverse populations "know" in different ways, as well as with commonalties between ways of knowing between populations? Ecopedagogical work problematizes our ways of knowing (i.e., epistemological frameworks) the world as well as how we understand and act toward the rest of Earth, deconstructing not only socio-environmental ignorance in teaching but also the things in differing epistemologies that allow for such ignorance.

Although learned epistemological frameworks can either, to varying degrees, connect or disconnect us from the rest of Earth, these complex connections exist. In other words, the subjectivity of the world does not "disconnect" us from the rest of Earth, no matter how "inconvenient."[3] Ecopedagogical work focuses on trying to understand these complexities, through diverse knowledges and sciences, to determine how we can better understand ourselves (i.e., humans) as part of the planetary sphere. This includes determining epistemological and knowledge limitations that separate rather than connect us with the rest of Earth. This book will discuss various possibilities of widening understandings that help to better understand the connection within and between these spheres (and citizenships) to help better guide our actions upon the rest of Nature, including teaching through epistemologies of the South (Figueroa and Harding, 2003; de Sousa Santos, 2018; Harding, 1998), citizenships from the local to the planetary, problematizing "development" and "sustainability," decoloniality, and decoloniality of globalizations (i.e., globalizations that can be correctly termed neocolonialism), among others.

This book will be theoretical but also practical. Although some readers might be frustrated that solutions will not be outrightly given, such as a step-by-step method or a linear path from "point A" to "point B," I argue that the vagueness is not a weakness, recognizing the complexities and need for contextuality through bottom-up praxis. Declaring a single pedagogy (or ecopedagogy) as the solution without local contextuality is a false declaration, with cookie-cutter pedagogies for everywhere resulting in problematic pedagogies everywhere. As expressed by Gadotti (2008a, 2008b) in the Earth Charter, ecopedagogies are only effective if democratic and contextual, with the ecopedagogical scholarship defining the grounding tenets rather than constructing static curricula.

At the heart of these arguments is bottom-up reinvention, which Freire argued was (2004, 1993, 1992, 1985, Freire and Macedo, 1996) essential for transformative pedagogies, as well as his own work. Raymond Morrow (2019) discussed how reinvention caused a "million Freires" from Edward Said's "traveling theory" and the connections to praxis, as dialectical education, "though criticized from neo-Marxist perspectives as 'idealist,' has had an immense practical impact on this transformative side of integral reproduction theory, precisely because of its capacity for historicization and nondogmatic reinvention." Sometimes, during his visits Freire would be asked for solutions for community(ies) problems, to which he would answer that he did not know, because he did not know their context. Rather, he would say that he could listen and give his own thoughts on what worked and did not in other places, but that he did not know what the solutions were without knowing their contexts. With the term "reinvention" Freire discussed the unfinishedness of humans and societies to keep the essence of the fluidness of social phenomena, including philosophies, theories, and pedagogies, as context differs and the present becomes the historic (i.e., time does not stand still, nor do societies). This is not only true for ecopedagogies but for all environmental pedagogies. As will be discussed later, when environmental pedagogies remain rigidly static, it is often more problematic than not having environmental teaching in the first place.

Ecopedagogical work does not define all these complex connections discussed until now, but rather problematizes the politics of epistemologies, knowledges, and sciences to determine the power dynamics between them locally to globally as affecting the planetary sphere (i.e., Earth overall). It is through such problematizing that necessary actions, including the location of those actions, to counter socio-environmental violence at its roots are determined and a better understanding of the effects from local-to-planetary levels is arrived at. Questioning and continuous re-questioning (or Freirean reading and rereading) of the politics within and between these complexities is essential to deconstruct what is taught as "legitimate" and "illegitimate" knowledges and sciences to understand the world, Earth, and the (dis)connections between them even though the world is part of Earth. However, the world's fluidness and subjectivity in determining how we comprehend the rest of Earth do not alter the environmental effects determined by the laws of Nature (Misiaszek and Torres, 2019; Misiaszek, 2018b). Worded more bluntly, Earth will return to a state of balance and continue to "exist" but the state of, or even existence of, the world within Earth is not guaranteed.

Ecopedagogical work is both deepening and widening understandings for praxis toward balance with the rest of Earth and socio-environmental peace for the world. Reading the world locally to globally, as part of Earth—as part of the planetary sphere—is the essence of ecopedagogical work. Ecopedagogical work should widen our world as part of Earth, with our actions in the name of "development" problematized within the planetary sphere. Such planetary perspectives are widened from critical global perspectives in which we act for socio-environmental justice for all the world—inclusive of all human beings. But it also includes the need for deepened understandings of locally contextualized perspectives.

Without needing to be stated, the complications of, and dissimilarities between, local-to-planetary perspectives of well-being, both within and beyond the anthropocentric sphere of the world, are immense. A key but difficult question is how the world's diverse contexts, including differing epistemologies, can have possibilities of praxis for global collectiveness toward all-inclusive socio-environmental justice and planetary sustainability, while being locally meaningful. The multiple, diverse epistemological lenses necessary to achieve these planetary goals will be discussed throughout this book. Ecopedagogical work includes the formidable task toward these goals, especially in connecting ourselves to the rest of Nature, which many epistemologies, such as the majority of the dominant Western ones, distance us from the non-anthropocentric sphere (Figueroa and Harding, 2003; de Sousa Santos, 2018; Harding, 1998).

Widened teaching of the vastness of the causes and effects of environmental violence to the global and planetary spheres can only be effective if it is accompanied *with* teaching that understands the local context and local epistemological framings of the violence (i.e., deepened teaching) (Misiaszek, 2018b). In addition, there is a need to question one's own epistemologies that frequently justify environmental violence, which can often only be done through other epistemological framings that problematize "how we know what we know." I will argue later that ecopedagogical work is essential for such problematizing within and, more importantly, outside traditional learning spaces. Ecopedagogies counter the furthering entrenchment of epistemologies that accept socio-environmental violence and unsustainability without question, resistance, or problem-posing if there are alternative ways of knowing (as well as problematizing why the term "alternative" is seen as appropriate here). There are various reasons for teaching through widened and depended perspectives, epistemologies, and pedagogies that will be discussed further in this book, along with an important reasoning by the Freirean notion that it is through one's own understandings, along with others as subjective beings, that transformative learning can happen. The "widening" of socio-environmental knowledges, including the connectivity with the non-anthropocentric sphere, should be taught (and read and researched) through one's own understandings and positionality—through their own epistemologies on how they understand themselves, the world, and Earth.

Freire (2000) used the term "thematic universe," which will be discussed more in Chapter 3, as "the reality which mediates men, and to the perception of that reality held by educators and people, that we must go to find the program content of education" (i.e., generative themes) (2000). Understanding through other persons' thematic universes is essential within ecopedagogical work to better understand, and problematize, our own self-reflectivity, how environmental violence is self-interpreted as compared to how others understand and are impacted by environmental violence. So is, although in many ways more difficult, understanding environmental violence through nonhuman "perspectives" to comprehend Earth's domination by "us" and "them." Ecopedagogies, as well as critical pedagogies overall, teach to read environmental violence through others' thematic universes, with specific emphasis on reading through othered thematic universes who suffer the most (Gadotti, 1996). There are numerous complications of such reading that will be discussed more later in the book, including the recognition of limitations of positionality to understand others' thematic universe (e.g., issues raised with feminist standpoint theories) and difficulties in determining whose suffering is essential to understand (e.g., argument of Boaventura de Sousa Santos' abyssal line positioning).

Throughout this book the term "ecopedagogical spaces" refers to teaching, reading, and research, and can be any type of education (formal, informal, and nonformal spaces), as well as individual spaces. Emergent from and grounded in Freirean pedagogies, ecopedagogical spaces of teaching, reading, and research are democratic and dialectic. Both of these characteristics are complex and will be discussed through the book, but being democratic is to meaningfully include all the knowledges, experiences, and epistemologies of those within learning spaces. Being dialogical is to construct learning spaces for all the teacher's(s') and students' knowledges to be part of the space's teaching and learning.

Some educational scholars argue against critical and/or Freirean approaches, but they misunderstand these approaches as nonstructured and undisciplined, relying on teaching from students' knowledges within the physical learning space, thereby severely limiting what can be taught. I have heard this criticism many times from numerous students and colleagues. It is important to note that being dialectic is not only to discuss within ecopedagogical spaces but also to discuss with those represented outside the space, bringing in voices through various resources, including the curricula (formal or informal or nonformal), within the ecopedagogical work at hand. In other words, the voices of those within ecopedagogical spaces are essential, but it is also essential to bring in diverse voices from outside. And as Michael Apple (2004) and de Sousa Santos (2018), among others, stressed, it is equally or even more essential to determine what voices and epistemologies are absent, so that they can be brought into pedagogical spaces. Ecopedagogical research involves deciding on the aspects of selection by the researcher(s)—whose voices will be selected, which methodologies will be selected, and what theoretical/epistemological lenses will be utilized. Within more individualized reading, ecopedagogical literacy is done through diverse perspectives and voices even if physically alone in the ecopedagogical space (e.g., ecopedagogical reading).

1.1 Dialogue: Critical and authentic

Democratic dialogue that grounds ecopedagogical work will be discussed throughout this book, with the discussion in this section provided as just a brief introduction. Ecopedagogical work is through dialogue from multiple, diverse cultural perspectives within learning spaces (i.e., multicultural learning spaces), as well as bringing in the world's cultural diversity within spaces of teaching, reading, and research not directly represented in the learning spaces. Thus, countering environmental pedagogical work narrowed through only one or one set of knowledge(s) and epistemology(ies) to understand the world-Earth thins "knowing" and decision-making (non-praxis) which can be truly devastating for the world-Earth.

Some scholars have argued that critical/Freirean pedagogies take away from the authority of the teacher. Although critical pedagogies do inherently dissolve the teacher-student hierarchies, teachers still have authority in learning spaces, especially formal learning spaces, as well as the responsibility to teach the specific lessons at hand.

> Authentic authority is not affirmed as such by a mere transfer of power, but through delegation or in sympathetic adherence. If authority is merely transferred from one group to another, or is imposed upon the majority, it degenerates into authoritarianism. (Freire, 2000)

Ecopedagogical spaces aren't ones that take up a "free for all" format in that any discussions and topics will be discussed; rather, teacher(s) facilitate dialogue through the students' diversity of understandings and experiences toward diverse understandings of the environmental topics at hand. Teaching toward a subject is not static—which could, in many ways, be defined as authoritarian—but rather forms spaces of thick, diverse dialogue for learning the topics at hand, with the recognition that conflict will almost certainly emerge.

Freire (2000, 1998b, c, 1997) has argued extensively that all education is political and that neutral education is not only impossible but dehumanizing because our (i.e., the world's) vast diversity, with our abilities to dream and to have diverse histories, means we will be in conflict with one another. To reemphasize again, this conflict is within concepts of peace for ecopedagogical spaces needing to be safe learning spaces so that conflict does not lead to violence (e.g., physical, mental, sexual violence) that would be opposite of democratic, dialogical learning spaces for socio-environmental peace (Bajaj, 2008; Harris and Morrison, 2003). Overall, I am stating that these spaces are not without teacher's(s') authority and/or "undisciplined," but rather fluidly structured with respectful listening to one another and the valuing of each other's understandings and arguments. The teachers' authority is not lost within ecopedagogical spaces; rather, what is lost is the unwarranted oppressive authority that silences students' voices and devalues their knowledges when they enter through the tangible or metaphorical classroom door. What is lost is what Freire (2000) termed as "banking-education" models that give sole voice and unbounded authority to the teacher (or "educational source") with her/his/

their voice as the only one from the only "true" source of knowledge and all others are silenced/dismissed.

There are various reasons for the need of democratic, dialectic environmental pedagogies for ending oppressions within and outside learning spaces, including the fact that all learning spaces are ideological spaces. This is especially true for formal education spaces. As Michael Apple (2004) has stated, critical ideological analysis to deconstruct hegemony is essential to problem-pose how "institutions of cultural preservation and distribution, like schools, create and recreate forms of consciousness, that enable social control to be maintained without the necessity of dominant groups having to reason to oven mechanisms of domination." As often an ideological apparatus of the State, with politics of powerful entries that cause environmental violence, there is a key question that emerged as a theme from my ecopedagogical research in the Americas (Misiaszek, 2011): Can ecopedagogies, as radical pedagogies, be truly part of formal schooling that questions and often counters dominant ideologies of the schools in which they are situated in? The silencing of students' voices within banking-education models effectively sustains and often intensifies dominant ideologies, as dehumanizing in such teaching methods as in its real-world results, and causes devastation to Earth (i.e., "deplanetarizing," a term discussed in more detail later). But what happens when such dominant ideologies are entrenched, as they often are, within schooling? An example of this is the ideologies that (in)directly justify unsustainable socio-environmental violence by striking down arguments that go against such unquestioned ideologies. When set within such ideologies, injustice and dominance continue; they lack the education that teaches possibilities and the need to protest and seek transformation. In contrast, critical models are inherent enemies of authoritarianism and dominance, including societal ideologies, especially neoliberalism.

Democracy and dialogue are essential within learning spaces for ending injustices and unsustainability outside of the spaces—in "local" societies within the global sphere. Such critical environmental teaching is essential to challenging and ending such oppressions and dominance, by determining what changes are needed and the obstacles in achieving them—Freire (2000) defined them as "limit situations" (see Chapter 3 for more discussion on this). Central to this is the continuous reading and rereading of the politics of socio-environmental violence. Teaching through democratic dialogue helps determine limit situations that prevent the ending of such socio-environmental injustices and domination, to then problem-pose possible solutions.

1.2 Ecopedagogies: Plural and intersectional

As stated in the introduction, ecopedagogies are not necessarily separate environmental pedagogies; they can also be transformational inclusions of other environmental pedagogies. Ecopedagogy can be both a separate environmental pedagogy and a pedagogical tool within Education for Sustainable Development (ESD) and Environmental Education (EE), as well as a methodological research tool for critically analyzing environmental pedagogies, including "pedagogies on the environment" (Misiaszek, 2018b).[4]

There is the argument of why we need to have environmental pedagogies, including ecopedagogies, if education for social justice inherently includes education for environmental justice, arguing possibly that the designation of "environmental" pedagogies further separates the connections between social and environmental injustices. In addition, there is the question of whether the "ecopedagogy" designation belittles previous critical work within ESD and EE. These arguments coincide with critiques of ecofeminism that merge feminism and environmentalism by often ignoring past feminist work that countered environmental exploitation and domination ((Lahar, 1991) citing (Cameron, 1989)) (Alaimo, 2008). Throughout this book, I will argue that ecopedagogies' focus on the sociohistorical politics of environmental violence calls for separate ecopedagogical work, inside and outside of social justice pedagogies, as well as other environmental pedagogies.

Ecopedagogical work is not to belittle the critical environmental work within other environmental and/or social justice pedagogies, but rather to center environmental exploitation and domination as critical theories center specific sociohistorical oppressions with the realization of their intersectionality with other oppressive types of othering. An example would be the intersectionalities of feminism and critical race theories.[5] Coinciding with the inclusion of ecofeminism, Lahar (1991), among other feminist scholars, has argued that the inclusion of new theories (and pedagogies, I would include) cannot emerge ahistorically without the recognition of others' past hard work and sacrifices.

In addition, it is essential that ecopedagogical work is not singular in framing (i.e., a single static ecopedagogy), a movement, or a philosophy in teaching, but rather plural ecopedagogies. This also coincides with ecofeminism as not singular in framing or with specific goals, but rather as having underlying goals of the intersectionalities of feminism and environmentalism that are contextually diverse (Alaimo, 2008; Lahar, 1991; Twine, 2001; Warren, 2000, 1994; Warren and Cheney, 1991). The intersectionalities of ecopedagogies with other socio-environmental justice-based pedagogies, including EE and ESD, are essential. I have attempted to explain this through pedagogies grounded in ecopedagogies; however, these arguments could be further deconstructed with pedagogies grounded in critical race theories, queer theories, and critical media culture, among many others focusing on teaching to end oppressions.

1.3 World-Earth: Complexities and conundrums

There are various complexities between "the world" which includes all human populations (i.e., anthropocentric sphere) and Earth (i.e., planetary sphere) within environmental pedagogies (please refer to Figure 1.1 while reading this section). One complexity, and I would state also a conundrum in many epistemological framings, is the world understood as within (or part of) Earth. My previous research has found that many environmental pedagogies structurally teach the world as separate from the rest of Nature (Misiaszek, 2011, 2016b, 2018b). Beyond ideological connections with the rest of Earth, there is also the very tangible and "practical" aspects that Ivan Illich (1983) expressed in the last chapter of *Deschooling Society*, stating that human limits and balance with Nature have been ignored in the endless pursuit of our "wants" (and "wants" perceived as "needs") constructed by scientists, engineers, and planners. Rather than perceiving humans as part of Earth, pedagogies on the environment often separate the world, as well as giving "us" the role of dominating the rest of Earth. This separation can also be seen in the terms "oppressing," used for within the world, and "dominating," used for the rest of Earth. It is essential to note here that although the wording will not be entirely precise throughout this book, the term "dominated" is more appropriate than "oppressed" for the non-anthropocentric sphere. This is because the non-anthropocentric sphere has no counter actions to humans' acts of environmental violence beyond automatic mechanisms to return to balance (i.e., equilibrium) dictated by the objective laws of Nature rather than being subjectively determined (Warren, 2000). Oppressions emerge when such self-determination is denied by others.

Problematized here in this section is how we are often taught world-Earth separation that leads to socio-environmental "distancing" that devalues all that is outside the world. I have and will continue to utilize the term "world-Earth" to designate this book's discussions on this subject.

Referencing back to Figure 1.1, the (dis)connectivity between the world and Earth would be taking the world (i.e., anthropocentric sphere) outside of the biocentric and planetary spheres. In one of my previous books (2018b), I used the wording of "Anthropo-Bio-Centric Gaps," as depicted on the graphic, with the aspects of distancing the world from being part of Earth. Although I won't be utilizing this term in this book, I would use the terminology that I used more in my earlier work on "distancing" (2011), with the term "world-Earth distancing" as essentially the same. Though my previous research (2011, 2016b, 2018b), I have argued that many environmental pedagogies distance us from the rest of Earth and that the distancing is intensifying along with globalizations from above (a topic discussed more in the next chapter and throughout this book).

Teaching that distances (e.g., geographically, time-wise, and othering)[6] environmental violence sustains/intensifies those actions by hiding associated social violence (i.e., hidden curricula for justifying environmental violence, and associated social violence, to continue) and unsustainability, thus leading to the lessening, or outright removal of, resistance for environmentally violent acts. Distancing narrows knowledges and understandings of socio-environmental connections and devalues the non-anthropocentric sphere as separate from the world. Figure 1.1 illustrates the aspect that "the world" is part of Earth, although many epistemological framings have the world as strictly separate from the rest of Earth. This is especially true for many Western epistemologies and/or epistemologies of the North. I argue within ecopedagogical framings that such false separations, when not epistemologically problematized in critical teaching and reading, will only sustain and intensify unstable socio-environmental violence. Distancing also refers to misunderstandings about who reaps the benefits of environmental violence in the name of "development." Distancing is also further exponentially increased through post-truth*ism* currently—see Chapter 7 on post-truthism. The term distancing will be utilized and described further throughout this book.

Ecopedagogical work inherently has the goal of de-distancing (or narrowing the Anthropo-Bio-Centric Gaps (see (Misiaszek, 2018b))) for understanding "our" world as part of Earth through countering epistemological framings that further distance the falsely termed "two" as rather "one." world-Earth de-distancing is largely what Freire wrote on ecopedagogy in his later writings. Along with Carlos Alberto Torres as a second author, I humbly wrote a hypothetical fifth chapter to Freire's book *Pedagogy of the Oppressed* (2000), entitled "Ecopedagogy: The Missing Chapter of *Pedagogy of the Oppressed*" (Misiaszek and Torres, 2019). No doubt controversial, especially to those who view Freire's most famous book as almost a religious text with any attempted additions or alterations as blasphemy, our chapter gives both descriptions of and justifications for how Freire discussed ecopedagogy to be the central theme of his next publication, unfinished due to his untimely death. In our chapter's second half, we attempted, once again humbly, to write the "Missing Chapter" to hypothetically be the additional chapter of Freire's book with numerous footnotes of Freire's own writings and discussions, including discussions with Carlos. I will discuss this work further in this book's third chapter on the Freirean reinventions of ecopedagogies, but in this section I want to refer to what we grounded our fifth chapter of his most famous book on, by problematizing real and false distinctions between "the world" and "Earth," as well as distancing between the two.

It was important to treat world-Earth differentiation as a foundation of this "missing chapter" because the differentiation allowed us to emphasize how Freire (2000) discussed the world as people, collectively and democratically, and how he constructed the world, with there being no world without humans (and diverse populations of humans). This is with the important acknowledgment of unfinishedness with human/societal transformability from "our" unceasing historical reflectivity and dreaming of possible futures (Freire, 2000)[7]. Education that does not allow for such reflectivity and dreaming is dehumanizingly fatalistic (Freire, 2000).

The unfinishedness and utopic aspects of the subjective world within the rest of Earth guided by the laws of Nature will be discussed further in Chapter 3. Ecopedagogical teaching is done by deconstructing the foundation of environmental teaching emergent from sociohistorical constructions of oppressions—which many critical scholars argue specifically (Apple and Au, 2009; Held, 1980; Gadotti, 1996) is essential for any pedagogy. I would include the ecopedagogical need for critical deconstruction of sociohistorical world-Earth distancing that politically sustains and intensifies socio-environmental oppressions, dominance, and unsustainability.

In "Ecopedagogy: The Missing Chapter of *Pedagogy of the Oppressed*" (2019), Carlos and I introduced Freire's (2000) following quote to ground this argument.

> A peasant who by banking standards was completely ignorant said: "Now I see that without man there is no world." When the educator responded: "Let's say, for the sake of argument, that all the men on earth were to die, but that the earth itself remained, together with trees, birds, animals, rivers, seas, the stars . . . wouldn't all this be a world?" "Oh no," the peasant replied emphatically. "There would be no one to say: 'This is a world.'"

Freire (2000) discussed how humans are the only beings that are self-reflective, historical, and able to dream of possible utopias; however, environmental pedagogies often ignore these aspects of the world as being non-dialogical, nondemocratic banking-education models that are fatalistic. The ecopedagogical question is how best to teach about the rest of Earth as understood in terms of both affecting the world and Earth's "non-world" outside the anthropocentric sphere. This non-world is the space, or the non-anthropocentric sphere between the world and Earth, that has been devalued in connection with the world and much more devalued in importance outside of those aspects that affect humans.

It is ecopedagogically essential to problematize how we teach to understand, care about, and care for this non-anthropocentric sphere, with the recognition of its existence beyond that of how we "understand" the non-anthropocentric sphere, as well as realizing the importance of what is "not known" beyond our current and future understandings. The dilemma of education within the world to "be" with all of Earth is a complex one within ecopedagogy that has no specific answers, but rather possibilities of learning that recognize the limitless complexities and the limitations of our understandings of all that is nonhuman. As a quick note, this is also true in the limitations of holistically understanding the world but is intensified outside of the world.

There are many framings and terminologies indicating such world-Earth limitations of knowing and the resulting distancing, (i.e., widening the Anthropo-Bio-Centric gap) as mentioned previously, with the following description.

> Ecopedagogues teach to narrow and ultimately remove the gap [(i.e., Anthropo-Bio-Centric gap)] . . . in which praxis for global justice can be achieved, and understood as only achievable, through widened actions for planetary justice. With this, the Freirean limit situations to be discussed in ecopedagogical spaces are the socio-environmental barriers that widen the Anthropo-Bio-Centric gap for praxis to emerge that overcomes these barriers (i.e., narrowing towards ultimately removing the gap). (Misiaszek, 2018b)

In describing the "gap," I focused on problematizing how we are taught the widening distances to further devalue the rest of Earth, along with the devaluation of its degradation as something not affecting human populations. There exists almost limitless points of empowerment and oppressions within the world, with ecopedagogies focusing on how these points connect with the widening gaps with the rest of Earth as both causes and effects of oppressions within the world and dominance upon the rest of Earth (Misiaszek, 2018b).

This utopian ecopedagogical goal is to narrow this gap through better understandings of the non-anthropocentric sphere through diverse perspectives and epistemologies, with the recognition that this gap will always exist with the limitations discussed previously, as we are "focusing" in reference to the previous sentence of "limitless points." Aligned with the notion of utopias within Freirean critical work, such goals will likely never be fully achieved but rather are targets to strive toward in ecopedagogical work, as how António Teodoro and Torres (2007) exemplified utopian goals as the horizon and others use the navigational metaphor of the North Star. Freire (2000, 1992) discussed how teaching should allow for students to be able to dream of possible utopias, in opposition to banking models that are structurally fatalistic without transformation.

In the below quote, Freire (2000) describes this as dehumanizing as we are beings that dream, have histories, and reflect upon our actions, unlike all other Earth beings.

> Unable to decide for themselves, unable to objectify either themselves or their activity, lacking objectives which they themselves have set, living "submerged" in a world to which they can give no meaning, lacking a "tomorrow" and a "today" because they exist in an overwhelming present, animals are ahistorical. Their ahistorical life does not occur in the "world," taken in its strict meaning; for the animal, the world does not constitute a "not-I" which could set him apart as an "I." The human world, which is historical, serves as a mere prop for the "being in itself." Animals are not challenged by the configuration which confronts them; they are merely stimulated. Their life is not one of risk-taking, for they are not aware of taking risks. Risks are not challenges perceived upon reflection, but merely "noted" by the signs which indicate them; they accordingly do not require decision-making responses.

Some environmental pedagogical scholars point to passages like the one above to state that Freire was placing the world as valuable and the rest of Earth as valueless. I would strongly disagree with this and would state that Freire wrote many passages after *Pedagogy of the Oppressed* (2000) that emphasized this, some of which are throughout this book and the conceptualized "missing" Chapter Five I wrote with Carlos (2019), among other publications. Freire here is not positioning the non-anthropocentric sphere to be devalued and dominated, but rather there are humanizing factors of the world that are unique. Freire's essence of hope and utopia in pedagogies will be discussed in Chapter 3 and throughout this book, with this section emphasizing the unfulfilled-*ness* of utopic goals of all-inclusive understanding of the non-anthropocentric sphere and world-Earth connections. In other words, educational goals should be going toward holistically understanding the world-Earth but also, just as important, the limitations of such "understandings."

Such all-inclusive understanding of the world is also unachievable. The endless complexities of the world can be argued through Freire's (2000) writings on how everyone has his or her own thematic universe that coincides and conflicts with others' Freirean labeled "thematic universes," with the goal of determining possible "generative themes" for dialectic learning spaces. Understanding the non-anthropocentric sphere is even more complex because we don't "know" fully what it is to be not human. Ecopedagogical world-Earth work problematizes the interconnections of the world as it affects the non-anthropocentric sphere, including the (mis)understandings of the non-anthropocentric sphere affecting Earth holistically. The overall question is how these (mis)understandings of the non-anthropocentric sphere cause social violence, with importance "given," or lack of importance given, to how the world's populations view destruction beyond affecting the world (i.e., within the non-anthropocentric sphere outside of affecting the anthropocentric sphere).

Although understanding the rest of Earth can only be done through "our" (i.e., human) perspectives, I would argue that critically acknowledging these inherent limitations of knowing is essential for praxis. It is foundationally important in ecopedagogical work to rigorously problematize that we, as humans, understand the non-anthropocentric sphere within human lenses, and just importantly, that there are limitations to such understandings. Without such acknowledgment, environmental teaching, including ecopedagogical work, is inherently flawed. It is through critical comparisons of how we understand the world, as a human and with other humans/populations as social beings, that we understand the rest of Earth. Both the advances and limitations of understanding the non-anthropocentric space—as well as the recognition of increased suppressions within environmental banking-education spaces and systems, a topic further discussed in Chapter 3—must be problematized. In short, the goal of understanding the non-anthropocentric sphere is utopic, with the realization that we'll never reach the horizon or North Star, as here I am utilizing Teodoro and Torre's (2007) metaphor again.

Revealing these limitations as existing is not to disparage learners by suggesting that absolute understandings will never be achieved, but rather to teach through these limitations to problematize hubris. However, some entities exploit notions of incomplete understanding, such as through post-truthism, to pervert teaching of falsities as truths. Such teaching is falsely justified through the gaps in understanding the non-anthropocentric sphere to mean that constructed falsities could be truths because "we just don't know" (Misiaszek, 2020). Post-truthism will be discussed more in length later (see Chapter 7), while problematizing falsities generated by arguing that limitation of knowing the non-anthropocentric sphere justify them, with post-truthism as causing extreme cases of this outward teaching through lying. A good example is climate-change denial whose justification is that we "just don't know" everything about the climate and Nature. Such education on climate change does not have the goal of understanding truth but rather works toward teaching an ideology(ies) that systematically prevents actions that counter hegemony (e.g., profit and associated power of fossil fuel industries). There are endless socio-environmental questions to problematize stated "truths" to determine the politics of the "truths" of the non-anthropocentric sphere and asking if the politics are diverting "true" truths from the objective laws of Nature. Once again, it must be emphasized that our knowledges of the world are dependent on our subjective interpretation of these laws, so the "true" truths are understood within "our" subjectivity.

Utilizing Durkheim's work, such knowing calls for comparative studies rather than direct experimentation, because it is through our interpretation of direct experimentation that we construct the world's understandings (Durkheim and Lukes, 1982) and our understandings for the rest of Earth. For example, we can determine the temperature changes of the planet, but how we understand and utilize such data in determining our actions is subjective and needs critical comparative analysis. Giving the example of education for development, ecopedagogical work, inside or outside ESD models, must problematize how we use our knowledges of the non-anthropocentric sphere to define "development." Is "development" determined within the balance between the needs (and "wants" within social justice models) of the world and Earth, or countering such goals of sustainability and socio-environmental justice? Taking the example of climate change, is this factored in when "development" is defined and framed? Ecopedagogical work through problem-posing these and similar questions of "development" and climate change, with/out hubris of acknowledging limitations, is the goal of reading to understand Earth, the world, and the dependent connections between them.

So, within all these endless complications of understanding the world and Earth with the existence of inherent limitations to our absolute understanding, is there an innate frustration in ecopedagogical work? I would answer "yes," at times, but also "no" (somewhat) in that realizing what we don't know and striving toward knowing "that" (and to repeat this process endlessly) is essential in all educational work, including ecopedagogical work and overall "development." Better world-Earth understanding is best accomplished through widened ways of knowing. Such widening is emphasized throughout this book that calls for world-Earth understanding through multiple epistemologies, coinciding with the postcolonial work of Bonaventura de Sousa Santo (2018, 2016, 2012a) and

Raewyn Connell (2014, 2013, 2007), Red Pedagogical work of Sandra Grande (2009, 2004), and ecofeminism work of quilting by Karen Warren (1994, 2000), among many other scholars who emphasize the need for multiple ways of knowing. What is not possible and attributes to social injustice, according to these and other scholars, is having a singular epistemological lens that dismisses all others. This is especially the case when a dominant epistemological framing is forced upon population(s), such as was/is done through (neo)coloniality. De Sousa Santos (2016) has emphasized that if we want to go to the moon, Western epistemologies, sciences, and knowledges are necessary, but if we need to know about planetary diversity, diverse epistemologies are necessary, especially nondominant, non-Western ones. What is also essential is teaching through the epistemologies of those within the relative ecopedagogical learning spaces.

1.3.1 Organic Ecopedagogical Modeling (OEM)

As a previous, not-so-good, undergraduate student majoring in chemistry before changing to the discipline of environmental studies, I was always intrigued with modeling of organic chemistry, even though the organic chemistry courses I took made me realize that I did not want to pursue the field of chemistry. What intrigued me was the modeling used to illustrate organic chemical reactions—first on paper and then using a small plastic set which featured multi-plastic balls with holes in them that allowed connection with other balls via bendable rubber sticks at varying lengths. What this toy-like kit allowed for is to put together molecules to see, in a very physical and tangible way, how molecules are formed by determining how elements are attracted to one another (i.e., bond together) or repelled.

Such modeling with this physical kit did not help me achieve success in organic chemistry, as can be witnessed by this overly simplified description here which probably would dismay many chemists, but it did spark my interest in what the modeling did not show—reality, or truly "knowing" what the molecules look like. In other words, the tangible modeling was made up. The modeling does represent what happens with elements and molecules (the attractions forming bonds, the repulsions, and breaking of molecular bonds); however, the models do not physically portray what we know they "look like" because we don't know what they look like. They could look like the modeling, but we just don't know. But what is of most concern to most chemists is that, by utilizing the models, we know what will happen when mixing chemicals in various types of environments.

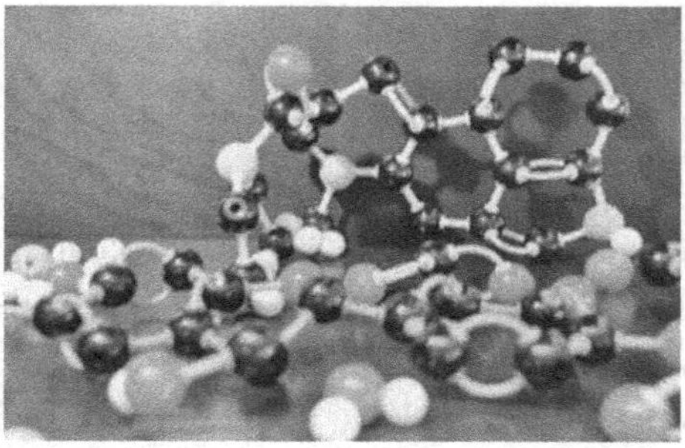

Figure 1.2 Photo of organic chemistry modeling kit. Photo credits: Greg William Misiaszek, 2019.

The reason I bring up this modeling is because I argue that understanding the non-anthropocentric sphere through diverse epistemologies can be metaphosphoric to organic chemistry modeling. Our own human experiences, knowledges, and understanding do not allow us to fully understand what the rest of Earth is, as "being" and "connectivity" (partially or holistically). However, we can teach and learn through diverse epistemologies to continuously understand the dynamics of such "being" and connectivity, with the continuous self-reflection that we could be getting it wrong in "reality" but gaining in understanding the non-anthropocentric sphere indirectly. In chemistry, predicting how chemicals will react with one another is of greatest importance, while knowing what they "look like" microscopically is given lesser importance.

In environmentalism and ecopedagogy, what is most important to understand is our actions as affecting the rest of Earth, as well as how it always rebounds to affect the world. Us, as humans, not being able to entirely understand the non-anthropocentric sphere, has us rely on different ways of knowing to understand this space, critically comparing epistemologies. An example is striving to fully understand the non-anthropocentric sphere's intrinsic value beyond our anthropocentric sphere. Beyond our world, our epistemologies cannot fully recognize all aspects of this, but there are epistemological framings that can help us toward this goal, as well as acknowledging our limitations of reaching it. The likelihood of sciences reaching the lesser goal of actually observing what elements and molecules tangibly look like is exponentially more likely to happen than us fully understanding the non-anthropocentric sphere. It is the widening of epistemological understandings that is essential in ecopedagogical work for Earth-world understanding, as well as countering political narrowing of epistemological teaching which is most essential.

Returning back to the concept of chemical modeling, this theoretical modeling without knowing all of what exists within the non-anthropocentric sphere is, as I term it, "Organic Ecopedagogical Modeling" (OEM). OEM is the epistemological modeling within teaching and research to utilize multiple, diverse *k*/*K*nowledges, *s*/*S*ciences, and *t*/*T*echnologies to conceptualize the non-anthropocentric sphere. The lowercasing and uppercasing of the first letters of these words will be discussed in later chapters in detail, but a simplified explanation of their usage is the notation of dominant forms for uppercased terms (i.e., solely from the Global North, within "epistemologies of the North" (à la de Sousa Santos) (2018, 2016)) versus nondominant forms for lowercased terms (e.g., Indigenous, localized, inclusive of the Global South, with "epistemologies of the South" (à la de Sousa Santos)).

With the key ecopedagogical goal of understanding the politics and effects of environmental violence from the world, OEM modeling helps us understand how human actions affect the non-anthropocentric sphere, as it inherently effects our world. As discussed previously, the latter part of this statement indicates the possibilities of achieving progression toward this goal but with recognition that all-inclusive and full understanding of the non-anthropocentric sphere is unachievable. This "uncertainty" does not mean inaction toward achieving it but rather, as discussed previously, the recognition that we are not part of the non-anthropocentric sphere. It is a space that we can observe and utilize un/sustainability, but to be within the non-anthropocentric sphere, to entirely understand and appreciate the non-anthropocentric sphere, is impossible beyond our interactions within it and observations of it. What OEM illustrates is theoretical modeling of diverse epistemological lenses to best know *within* the limits of our understandings outside the non-anthropocentric sphere.

Ecopedagogies' grounding of Freirean dialogue and democratic education to center work to end socio-environmental violence and unsustainability grounds the epistemological selection processes, which will be further discussed in Chapter 4 but also throughout this book. As OEM focuses on modeling that helps to understand the reactions of elements and molecules, OEM within ecopedagogies focus on what is needed to better understand the non-anthropocentric sphere for praxis of the world toward sustainability of Earth holistically and socio-environmental justice within the world. This coincides with the ecopedagogical goal of praxis of the world, with us as the only reflective unsustainable beings, the required transformation emerges from the world, through our own understandings, however limiting they are. The ecopedagogical goal of better understanding socio-environmental violence through multiple epistemologies is ultimately for deepened self-reflective praxis to end unsustainability where we *can* and *should* focus our actions. The "can" has already been initially described and the "should" indicates the negative consequences of altering natural processes (with recognition of the "butterfly effect").

OEM epistemological selections are endlessly complicated and difficult; with this recognition, this book will offer some groundings and possible answers, especially in the sections dedicated to epistemological discussions (see Chapters 4 and 5). This is with the recognition that the selection processes are inherently fluid and subjective with no absolute answers or easily followed step-by-step procedures. Ecopedagogical dialogue and democracy are essential in the selection processes as they are contextually determined through the world's diverse epistemologies and conflictive histories—almost as limiting as they are to truly comprehend the non-anthropocentric sphere. How we interpret the world-Earth is ultimately from us. Returning to the example of the chemistry modeling, chemists don't know what is happening in real-world mechanics, but they realize how this modeling connects back to real-world observations through our five senses. The model, which is fictional as interpreting the mechanics, is very real in the results of knowing chemical reactions.

With epistemological selections, it is the outcome of such understanding that determines our actions, which are either environmentally positive or environmentally negative for Earth holistically. If our actions are the ones needed to adapt for Earth balance and we "understand" only though our epistemological lenses, the selections should be pragmatic toward this end. Selections must emerge from our own histories and observations through democratic dialogue and critical reading (e.g., ecopedagogical literacies), which will be the focus of the next chapter.

Notes

1 Apple discussed this continuum of environmental violence in a lecture series at Beijing Normal University (BNU), China, in September 2013 and November 2019.

2 I write this sentence recognizing that the term "who" is limiting in English, as well as in many other languages, which calls for linguistic problematizing within ecopedagogical work also, especially with English as the most prominent lingua franca (which, in itself, must also be problematized in critical pedagogies including ecopedagogies).

3 This wording is playing off the title of Al Gore's *An Inconvenient Truth: A Global Warning,* 2006. Directed by GORE, A. Hollywood, CA: Paramount. Documentary entitled "Inconvenient Truths."

4 The term *pedagogies on the environment* denotes teaching that includes antienvironmental education and *environmental pedagogies* focused on environmental well-being (e.g., ESD, EE, ecopedagogy) Misiaszek (2018b).

5 "Ecofeminist analysis includes the human exploitation of the nonhuman environment in its list of interwoven forms of oppression such as sexism and heterosexism, racism and ethnocentrism Lahar" (1991).

6 I have further defined distancing below with an example of citizenships:

> I [have] found the aspect of distancing of social oppressions due to environmentally harmful acts, whether it be physical (i.e., corporations placing environmentally harmful production in developing countries), historical (i.e., as being inactive on climate change because the effect won't be felt for a "long time"), or aspects of othering (i.e., justifying environmental ills onto specific peoples seen as "not like me, not my fellow citizens"). Distancing [in this book questions] how can citizenship education counter othering to not "export" socio-environmental problems to other humans, to future humans not born yet, and to Earth as a citizen? (Misiaszek, 2018b)

7 Futures is in plural to indicate the utopic idea of possible multiple futures, rather than a fatalistically determined future.

2

Ecopedagogical Literacy

Reading the World within Earth

What does not seem possible to me is to read the word without a connection to the learner's reading of the world. That is why, for me, the literacy process with adults necessarily implies the critical development of the reading of the world, which is a political, awareness-generating task. What would be wrong, and what I have never suggested should be done, is to deny learners their right to literacy because of the necessary politicization there would not be time for literacy in the strict sense of the term. Literacy involves not just the reading of the word, but also the reading of the world.

(Freire, 1993)

Ecopedagogical reading involves reinventing Paulo Freire's reading of the world by expanding it to include reading Earth holistically, with the world as part of Earth (Misiaszek and Torres, 2019). Reading the world as part of Earth, as a "political, awareness-generating task," seeks to determine how all politics, ranging from the local to the global, are both the causes and the effects of violence against the world and the unsustainable dominance over the non-anthropocentric sphere, with falsely "taught" separations between the world and Earth. This planetary widening is not absent from Freire's works, especially in his later writings (Freire, 2004, 1997, 1993)—some passages from his works which exemplify this have been quoted throughout this book. Environmental pedagogies that systematically deny students readings of the politics of environmental violence as inherently dehumanizing and deplanetarizing are categorical traits of banking environmental pedagogies. Within banking models, reading is taught as technical within only the prescribed ideologies without scope for students' critical reflection, dialogue with others, or previous knowledges or epistemological framing(s), among other aspects of humanization. It is not impossible for students within banking models to bring about the necessary socio-environmental transformation, but banking pedagogies systematically devalue students' praxis outside of static ideologies and deny socio-environmental praxis within learning spaces. Thus, in banking spaces, the reproduction of inherently oppressive/unsustainable ideologies are not countered.

As discussed in the previous chapter, it is impossible to understand all the complexities of any social phenomena connected with Earth; in ecopedagogy, it is most important to teach to critically problematize environmental violence through various local-to-global lenses, with endless questioning requiring further problematizing. This reading of the world-Earth is ecopedagogical literacy (Misiaszek, 2020, 2018b). Ecopedagogical reading problematizes how our ways of knowing and others' ways of knowing influence socio-environmental actions of (non)violence and (un)sustainability, as well as the politics of the socio-environmental "knowing." Once again, it is important to remember that environmental violence is a continuum, so problematizing world-Earth sustainably is essential to ecopedagogical reading. As I discuss the aspects of ecopedagogical reading, Freire's work on which this is grounded will be discussed briefly in this chapter, but more elaborately in Chapter 3. In this chapter, I discuss the overview of the goals and foundations of ecopedagogical literacy, and such reading will be exemplified and further detailed throughout the book.

The ultimate goal of ecopedagogy is its literacy of reading and rereading toward praxis of socio-environmental justice and sustainability. This differs from one environmental pedagogy to another, especially within banking models, in which the goal is maximum quantity of didactic environmental information "absorbed" by students (Misiaszek, 2018b, 2011). In other words, it is the quality of criticalness of reading "new" (to the reader) environmental information continuously bombarding learners outside of learning spaces that should be prioritized, rather than the quantitative amount instilled. The words "instilled" and "absorbed" are utilized in a banking-education sense in that knowledge is provided and its retention tested without assessing if learners can or cannot critically analyze new world-Earth information; thus possibilities of praxis are devalued, if valued at all.

The teaching of critical literacies of problem-posing socio-environmental issues centers effective problem-solving, rather than gaining absolute environmental knowledges, as the goal of ecopedagogy. I have written the following on this topic:

> Students (and teachers) will certainly have more overall knowledge of environmental issues; however, it is the ability to widen and deepen critical reading abilities of environmental issues, politics, and what are next issues

> necessary [in order] to learn to better determine their actions towards ending socio-environmental oppressions. (Misiaszek, 2018b)

Here ecopedagogical literacy is reading to deepen and widen understandings of socio-environmental violence *upon* and sometimes *with* others, along with theorizing beyond the world to the planetary sphere (i.e., the rest of Earth). The term "others" indicates populations outside of our self-defined "own" population(s) that are differently affected by environmental violence. The use of "with" indicates how understanding others' socio-environmental effects can allow us to better understand our own. How can we utilize theories to understand the world's differences and interconnectivities with the non-anthropocentric sphere as all part of Earth? Ecopedagogical reading is through theoretical lenses, especially critical theories, to deepen and widen our self-reflectivity as we encounter environmental violence.

As ecopedagogies are plural, as discussed in the previous chapter, there is no single, or a specific set of, ecopedagogical literacy(ies) with the goal of understanding socio-environmental oppression and dominance from those who suffer the most. Ecopedagogical reading includes bottom-up perspectives of social violence from environmental violence (and vice versa), with the aim of better understanding Nature's dynamics within and outside of the world. Also, ecopedagogical literacy isn't through a single discipline or a single epistemology of disciplines, but rather through reading transdisciplinary insights and problematizing limits of the disciplines themselves. This latter part will be discussed later in this chapter and throughout the book, with Boaventura de Sousa Santos' (2018) arguments of sociologies of absences and sociologies of emergences expanded upon in Chapter 5. Ecopedagogical reading must incorporate, once again, limitations of knowing, with the realization that expertise in everything through all epistemological frameworks is impossible. Ecopedagogical work, including ecopedagogical literacy, cannot be apolitical or static, but needs to be read through others' eyes with the help of theories. Such theorizing is with the realization that such perspectives are imperfectly subjective and cannot lead to complete understandings. There are various theories discussing this goal of positionality of others, with one of the most prominent being feminist standpoint theory as defined by Sue Clegg:

> "Standpoint" theory, which in its strongest version in feminism claimed that women, by virtue of their distinctive experiences and through the development of a feminist stance, could have insights that others could not (for example, (Hartsock, 1998); (Harding and Hintikka, 1983)). Truth therefore (depending

> on the version of standpoint theory adopted) was (partially) relativised to social group and made context dependent. (Clegg, 2011)

Ecopedagogical literacy is a self-reflective reading in that it requires continuous determination of the limitations of what "I" know, not "know," and the aspects that need to be learned next for deepened and widened understandings.

2.1 Countering objectivity and neutrality: An epistemological introduction

Ecopedagogical literacy coincides with critical literacy and teaching that centers on deconstructing our ways of knowing the world and Earth—the Freirean (2000) term of consciousization (i.e., *conscientização*). Ecopedagogical literacy endlessly problem-poses how we know what we know about the world-Earth, especially what we view as unquestionable (e.g., "common sense"). Are we taught, and do we "know" the world-Earth only through a single epistemological framing that is considered to be "objective," or through multiple, diverse epistemologies that are considered subjective? I have found in my previous research that many environmental pedagogies teach the former rather than the latter (Misiaszek, 2018b, 2016a, 2015, 2011).

Ecopedagogical reading and teaching is research—a statement that will be further argued later in this book. Utilizing the methodological question of whether research upon a fish's environment is better from inside (i.e., as the fish) or from outside the metaphorical "fishbowl," I have argued the need for teaching critical literacy that is research from <u>both</u> outside <u>and</u> inside the bowl (Misiaszek, 2018b).[1] Research and teaching from outside the bowl are essential, as only those outside can understand the contextual aspects of the fish, whereas the fish itself cannot fully compare and contrast its environment owing to its limited perspective from inside the bowl. But also, only the fish has the unique positionality to "know" its own environment. To read socio-environmental violence, reading must be through multiple, diverse epistemologies and individual/societal perspectives, especially from those who are most affected by the violence.

Many scholars have argued that the division between "individual" and "society" is not a binary one with clear divisions. For example, in the following quote, former prime minister and later president of Tanzania Julius Nyerere explains how we humans are inherently social beings.

> For man is a social animal. A man in isolation can be neither liberated nor educated; the words are meaningless in relation to an abandoned child brought up by wolves. And education is a social activity, with a social purpose. It is individuals who are educated. But they are educated by their fellows, for the common purposes of all members of the society. (Lema et al., 2004)

A critical question raised by this quote, beyond the problematic gendered writing, is whether our "education" for such roles and purposes in society has been narrowed by only epistemologies of certain societies and by ignoring others' epistemologies of knowing the world-Earth. There are a few aspects of the world that are absolute, positivistic "truths" across all populations holistically, but rather most "truths" are contextually subjective. This has been a topic discussed in the previous chapter in some detail and will be discussed throughout this book.

Ecopedagogical reading emerges from continuous self-reflection on these aspects of social subjectivity that are often difficult to comprehend because they frequently counter what is taught in schooling and how "it" is taught, especially in banking-education models. This includes environmental pedagogies that teach that "our" perspectives of world-Earth (i.e., our society's epistemologies as interpreted) are correct and "theirs" are incorrect because they conflict with "ours." Also, what could be difficult is problematizing our "instilled" epistemology(ies) because such revelations could counter our belief that we are free individuals with our "own" individual thoughts. Recognizing this and acting accordingly is overcoming various aspects of "fear," as Freire (2000) discussed how such consciousization counters our internalized, long-held positionality of one's Self within the world. Such consciousization needs deepened questioning of the politics of the "wide" world which instills critically reflective global self-positioning.

For example, individuals within Western societies that are ideologically entrenched with Western individualism will likely counter the aspect of individual thought as mostly socially constructed, touting individual freedom, as argued by Albert Scherr (2005) and other critical sociologists (among other disciplines). I would argue, with others, that the contexts of Western individualism, to differing degrees, are "spreading" through Westernization (as globalizations from above), largely accompanying with neoliberal globalization; these topics of globalizations will be discussed in the next section. With the ecopedagogical need for widened understandings of others' epistemologies and deepened understandings of others' perspectives, we deepen and widen our understandings of "our own" epistemologies through "their" ways of knowing. In other words, looking through one's own society through others' eyes allows for the possibilities of understanding one's own perspectives. It is through diverse epistemologies that better understanding and critical questioning of self-perspectives is possible. This is just an extremely brief discussion onto these aspects, with ecopedagogical literacies including "knowing" what is beyond the world as part Earth.

Ecopedagogical literacy is focused on trying to understand Earth holistically from diverse knowledges, as discussed previously, including through OEM modeling. Coinciding with these previous discussions, ecopedagogical literacy has unique characteristics distinct from other critical literacies in the non-anthropocentric sphere that have no "humanized perspective(s)" or epistemologies, or at the least we don't "know" of any. There is no perspective(s) outside the world, that is, to give nonhuman entities a "perspective" is likely to lead to falsely "humanizing" the non-world. It becomes extremely complicated because this aspect does not diminish the necessity of valuing the non-anthropocentric sphere in the same way that Freire did not devalue all other animals (Misiaszek and Torres, 2019; Misiaszek, 2011), as discussed previously. In addition, it is important to recognize that placing human characteristics upon nonhumans can also be problematic, with this taken as recognized limitations of our readings of the non-anthropocentric sphere.

With all these aspects in this section, the essence of ecopedagogical reading through the subjectivities of the world is essential. Ecopedagogical reading inherently counters false positivity (or false objectivity) of the world, as it is a part of Earth in which the non-anthropocentric sphere has the truly positivistic laws of Nature ("known" or not "known" by the world). It is reading through constant, unending reflection that all knowledges are political, and that full neutrality is impossible in the sciences, technologies, and the ways of teaching knowledges, thus there is a need to counter false acceptances of such knowledges as being apolitical (de Sousa Santos, 2018, 2012a; Freire, 2000, 1998c; Harding, 2006, 1991; Figueroa and Harding, 2003; Apple, 2004). Such false neutrality impedes upon understanding through diverse ways of knowing, including ecopedagogical reading of environmental violence. False neutrality also silences democratic dialogue by silencing others' perspectives, pedagogies, and epistemologies to read environmental violence.

De Sousa Santos (2018) expressed that "any system of knowledge is also a system of unknowns" in which the "unlearning" of knowledges is essential for "the flux of the different kinds of knowledge in which human beings are involved and engaged, ignorance can be seen both as a point of departure and as a point of arrival." An important ecopedagogical question is what knowledges and epistemologies are needed for ecopedagogical work, including literacy, for transformative praxis. I would answer this, acknowledging that there is no specific answer to this question, by returning to the OEM model in that it is the groundings and outcomes of the ways of knowing that lead to socio-environmental justice and world-Earth sustainability.

Epistemological selection, a topic discussed by many scholars (de Sousa Santos, 2018, 2012a; Harding, 2006, 1991; Figueroa and Harding, 2003; Connell et al., 2017; Connell, 2007), includes the need for critical reading of epistemologies themselves, with particular focus on their sociohistorical groundings. Deconstructing the histories of epistemologies requires reading on what our ways of knowing lead us toward and away from. For example, do our epistemological perspectives lead toward domination through the framings of Edward Said's Orientalism (1979), in which dominant colonial, Western lenses are only used to understand the world to help justify (neo)colonialism? De Sousa Santos (2018, 2016, 2007b) has discussed the epistemological abyssal line in the divisions between epistemologies of the South and epistemologies of the North, in which the working North/South is not indicative of geographic locations but of divisions of dominance. He explained this in the following statement: epistemologies of the South exist to counter the epistemologies of the North, which have developed through histories of capitalism, colonialism, and patriarchy, with the former epistemologies not existing without the latter's presence (2018, 2016, 2007b). Said and de Sousa Santo's postcolonial work will be discussed later at more length, but here it is important to note that false neutrality within epistemologies viewed as having absolutes of knowledges, perspectives, and pedagogies blocks all other ways of knowing, and unlearning becomes nonexistent.

There are various terms/phrases that express the need for world-Earth understandings to be multiple and for ways of knowing to be diverse. In this book, I will utilize de Sousa Santos' (2018, 2016, 2012a, 2007b) terminology of "ecologies of knowledges," with the goal of constructing "knowing" toward ending oppressions of the world and dominance over the rest of Earth. Ecopedagogical reading must be through ecologies of knowledges to better understand Earth, with critical questioning of one's own epistemologies and those of others toward, or away from, this goal. As stated previously, it is critical questioning and re-questioning one's own ways of understanding that is an essential step for ecopedagogical literacy. In particular, how do I understand the world and Earth, including the non-anthropocentric sphere, that leads to sustainability or unsustainability of environmental violence currently, historically, and possibilities of futures?

Utilizing the work of postcolonialism scholars, such as Memmi (1967), Fanon (1967, 1963), Said (1979), Connell (2014, 2007) and many others, we can, through problematizing the histories of domination and the epistemologies that viewed such domination over the non-anthropocentric sphere as positive and through self-reflecting, unlearn such false-positive justifications. Reading these histories to unveil hidden ideologies for domination is essential to understand current and future socio-environmental unsustainability. For example, what ways of knowing were necessary for justifying the colonial raping and exploitation of natural resources that caused social and environmental domination, dividing the world between the Global South and the Global North? How does colonial socio-environmental domination continue in the postcolonial world with the problematic mapping of "developing" and "developed" nations largely coinciding with mapping of past colonialism, with recognition of the problematic dualist language of denoting "development" to populations? To further this problematizing, why do the historical maps of colonialism and current mapping of the developed/developing world (or degrees of hegemony) coincide more frequently than not?

This leads to the question of the end of colonialism, and to that of how global capitalism reinvented colonialism as economic-based neocolonialism through neoliberal globalization. In the next section, the problematizing of globalizations that can be renamed, rather correctly, as "neocolonialism" will be discussed. The epistemological question within ecopedagogical work, including possibilities of OEM modeling for understanding the non-anthropocentric sphere, is this: What are the ways of knowing to help cease "knowing" that justify sustaining/intensifying dominance and unsustainability? In de Sousa Santos' terminology, this questioning becomes: What epistemologies of the South are essential to end dominance and unsustainability justified by epistemologies of the North?

Epistemologies of the North largely dominate by portraying false neutrality in that they exist within "sanity," "reason," and "logic" without political influence. I would argue that Northern ways of knowing are, unfortunately, very effective in continuing environmentally violent, unsustainable acts. For example, neutralizing the defining and goals of "development," with only one normalized framing within the measurements of capitalistic economics, ignores all other epistemological framings of "development" and other systems of economics. Through epistemologies of the North, "sciences" and "technologies" are largely viewed unquestionably as Western methodologies and tools developed through these methodologies. These are just some examples of problematizing that will be discussed later, in the next section on reading globalizations as affecting our reading of environmental violence.

Contrary to those of the North, epistemologies of the South are grounded on the view that "neutrality" "mak[es] no sense because the criterion for trust lies in the vicissitudes of the struggle against oppression, thus immediately precluding any contextual indifference" (de Sousa Santos, 2018). As within Freirean pedagogies, it is the grounding of teaching to better understand struggles from populations who most struggle that makes neutrality and total objectivity impossible, because it is relations between individuals and populations that cause struggles, violence, and inequalities. Social injustices from our violent actions upon Nature cannot be fully understood without knowing the contextuality and positionalities of those affected. Arguing with de Sousa Santos (2018), this can only be done thoroughly through ecologies of knowledges, especially through the epistemologies of those affected by the relevant violence.

Returning once again to the argument that full ecopedagogical understanding of the non-anthropocentric sphere is impossible, what is necessary is the utopic goal of understanding this sphere through multiple epistemologies toward knowing its limitations with endless reflectivity. Neutrality and absoluteness entrenched within epistemologies of the North not only cause false superiority over all other epistemologies but also devalue all other epistemologies, especially epistemologies of the South that counter them. In other words, if knowledges are viewed as "correct," "absolute," and apolitical, why would other ways of knowing be needed? Such epistemological devaluing is the barrier to socio-environmental justice and sustainability. As de Sousa Santos (2018) stressed throughout his work, it is a social injustice to understand only through a singular dominant epistemology, as well as a loss of knowledge for the world to understand Earth and ourselves within Earth. In short, inherent suppressions of understandings from epistemologies of the North counter world-Earth balance and social justice within the world.

2.1.1 Knowledges, sciences, and technologies

Continuing from the previous section's end, reading the sociohistorical groundings of knowing is essential, as knowledges, as well as the sciences that emerge from these knowledges, are contextual, historical, and political. However, there are various political processes that falsely instill that certain knowledges are universal, static, and apolitical. The former description is a contested terrain that can be empowering or oppressive, but the latter is hegemonic when the three characteristics are falsely satisfied. Coinciding with previous upper-/lowercase meanings of words, the capitalization of <u>S</u>ciences and <u>K</u>nowledges indicate dominant hegemonic framings (accompanied with *underlining and italicizing*) that reject all framings of sciences and knowledges. Coinciding with Sandra Harding's arguments (2006, 2004, 1998), such systematic rejection is inherent because hegemony relies on their absolute, unquestionable ideologies by instilling that there are no alternative ways of knowing to become "common sense." Ecopedagogical literacies are through multiple knowledges and sciences, countering socio-environmental violence and unsustainability inherent to *K*nowledges and *S*ciences, as well as outputs from them (e.g., understandings, *t*/*T*echnologies, methodologies, tangible objects).

It is important to note that there is no single science and knowledge (or set(s) of sciences and knowledges) that possesses the characteristics indicated through uppercased lettering (i.e., *S*ciences, *K*nowledges), but rather both the usages and sociohistories of them designate the essence indicated by the letter casing—this is a topic that will be discussed more throughout the book. An essential question to problematize is how knowledges and sciences have emerged. For example, dominant sciences are most often seen as Western sciences, but that does not always mean that they are *S*ciences. This example is complicated in that the "Western" does not indicate only practices and social systems within "the West," but rather that the sociohistories that emerged from Western epistemologies. For example the Western scientific method or economic models of capitalism, including neoliberalism, is Western; however, Japan largely incorporates them even though Japan is not geographically in "the West."

*S*ciences have the essence of objectivity coupled with universality, falsely grounding supremacy over all other ways of knowing

> entirely distinct from other possible justifications of trust, such as authority, consensus, tradition, revelation, or efficacy. Nonetheless, it is possible that trust in science may lie, at least in part, in these other justifications, even if disguised as objectivity (de Sousa Santos, 2018). While knowledges appropriate reality, ways of knowing embody reality. (de Sousa Santos, 2018)

As the second part of the quote expresses, ways of knowing "embody" all that "we" are as humans, perceiving each other and the rest of world-Earth. Constructing and reconstructing understandings of our realities, with education as a tool for acquiring knowledges to better understand the world and Earth,[2] calls for the need to problematize what knowledges are included, are excluded, and the politics of such selections. This coincides with the view that many critical scholars, such as Michael Apple and Wayne Au (Apple and Au, 2009; Apple et al., 2009; Apple, 2004), have emphasized: the role of critical pedagogues to problem-pose who has the "right" and/or "access" to specific knowledges. The connections of knowledges and ways of knowing are extremely complex, with the ecopedagogical need of teaching "legitimized" and beyond "legitimized" Knowledges to widen and deepen socio-environmental connections.

There are various ways to conceptualize this, including using the terminologies of renowned critical educational literacy scholar Robert J. Tierney, widening our "bandwidth" of understanding to widen our "global meaning making":

> Let us challenge the systems that confine us and are used to immobilize others. The bandwidth exists; the virtual architectural possibilities can be imagined. We have the capabilities to support different languages and epistemologies, various modes of thinking and expression, a range of different types of exchanges as well as afford multiple perspectives and to do so with ethical integrity and global relevance. (2017a)
>
> Increased access, heightened bandwidth, and a variety of outlets, digital spaces offer global meaning makers less-regulated venues to assemble, explore, share, and connect. (2018b)

Tierney's descriptions will be further explored later, but in this section we will discuss as to how we can widen, and deepen, understandings for ecopedagogical reading, teaching, and research. As ecopedagogues, the goal is to widen bandwidths of global meaning, with concerns of selection contextuality and necessary ways of knowing that expand our/their/others' connections with the rest of Earth, with particular attention to the non-anthropocentric sphere.

I have previously discussed de Sousa Santos' arguments of needing ecologies of knowledges and described my OEM modeling that might be helpful conceptually. Also helpful is de Sousa Santos' (2018, 2016, 2007b) work on epistemologies and the abyssal line, which provides a framework on problem-posing and the overall selection of epistemologies toward better world-Earth understandings. I do not have room to fully elaborate upon de Sousa Santos' concepts of the abyssal line; but briefly, one of the three factors influencing which side of the line epistemologies are located in is determined by their grounding or their countering of capitalism, coloniality, and patriarchy.

Ecopedagogical reading focuses upon how knowledges, sciences, and technologies have continued historical oppressions, such as coloniality leading to inequalities of socio-environmental violence in terms of degree, frequency, and impact of local-to-planetary unsustainability. Ecopedagogical literacy is reading to first deconstruct what are the knowledge foundations through which the Self, *with* others, understands socio-environmental violence. Utilizing the fishbowl analogy again, it is necessary to look from inside and from outside the "fishbowl" to recognize and deconstruct the epistemologies we use, especially subconsciously, on socio-environmental issues and Earth overall. For this task of understanding Earth beyond our limited knowledges of the world, ecologies of knowledges (á la de Sousa Santos) must be incorporated well beyond Sciences and Knowledges. It is well beyond problem-posing our own ways of knowing, but also to include others' ways of knowing to understand the world-Earth more holistically.

Before ending this section there are two important topics to be discussed—ecopedagogical work in constructing/reinventing new understandings and revisiting post-truthism in this section's topics. With regard to the first topic, constructing/reinventing understandings, our previous knowledges create new knowledges and sciences, as well as technologies, by reading the world through our curiosities, as well as respecting others' curiosities. Freire's quote below describes the aspect of how critical curiosities should continuously problem-pose how and what sciences and technologies "better" the world holistically.

> As a manifestation present in the vital experience, human curiosity has been historically and socially constructed and reconstructed. Precisely because the advancement of innocence toward criticalness does not take place automatically, one of the primordial tasks of progressive educational practice is exactly the development of critical curiosity never satisfied or docile. That is the curiosity with which we can defend ourselves from the "irrationality" resulting from certain excesses of our highly technological time's "rationality." This analysis, however, carries no falsely humanistic impetus against technology or science. On the contrary, it seeks to imbue technology with neither divine nor diabolic significance, but to look at it, or even observe it, in a critically curious manner. (Freire, 2004)

For ecopedagogical work, reading is widened to Earth with the world as a part, with curiosities that challenge rationalities of Sciences and Technologies that are dehumanizing and de-planetarizing (i.e., distancing us from the rest of Earth). Ecopedagogical foci include how we can teach for endless critical curiosities to problematize knowledges, sciences, and technologies in the various ways previously described. Such problematizing will likely lead to problematizing knowledges learned from banking environmental pedagogical models, as well as most pedagogies on the environment. How Freire argued that curiosities are the foundations and the guidance of research, including research in learning spaces, will be further discussed in Chapter 5.

The second topic that warrants discussion is the ecopedagogical need to deconstruct false Knowledges and Sciences constructed from intensifying post-truth*ism*, and the Technologies utilized for their legitimization and dissemination—topics discussed at length in this book's last chapter. The political practice of portraying falsities as facts is not new; however, the mechanics of it within globalizations, especially within communication technologies such as social media, is uniquely modern within our societies today (making the communication Technologies) (Misiaszek, 2020a,b, Misiaszek, 2020). These Technologies differ not only in their medium of distribution but also in their selection of the target audience. Ecopedagogical work, coinciding with critical pedagogies, problematizes all sciences, knowledges, and technologies as political; however, it is the political usage of opposing a path toward any sort of truth that makes post-truthism different from political manipulation of socio-environmental facts. This can be likened to what Harry Frankfurt (2005) termed as "bullshit[ing]," but this, again, is different from post-truthism (Misiaszek, 2020, Misiaszek, forthcoming, 2020). To differentiate in this book, I'm going to utilize the aspect that sciences and *k*nowledges have the foundational function of seeking truths, and technologies as countering Technologies' (de)legitimization and distribution. The latter part indicates the complexities of the same technology (e.g., the internet, social media)—that it can be both technology and Technology, depending on how it is utilized.

2.2 Globalizations: Contested terrains

Ecopedagogical reading deconstructs global processes to determine the positive and negative socio-environmental aspects within and between different spheres of the world within Earth holistically, which have increasingly fluid, overlapping, and blurred boundaries. Reading globalizations is essential for numerous reasons, but most important is that the effects of environmental violence do not respect geopolitical borders. For example, the nuclear fallout from the 2012 meltdown at the Fukushima Daiichi Nuclear Power Plant in Japan doesn't stop at Japan's borders. Global reading for determining actions for socio-environmental well-being is essential for local-to-global justice and local-to-planetary sustainability.

It is important to first note the plural terminology of globalizations that will be utilized throughout this book, coinciding with the scholarship of Carlos Alberto Torres (2009), indicating the contested terrain of the processes as contextually "positive" or "negative." Returning to the ecopedagogical foundation to better understand socio-environmental suffering from those who suffer the most, the key question for globalizations is as follows: How do the processes of globalization affect this violence, domination, and the sustainability between the social and the environment? And which populations suffer most through globalizations, both within populations and between populations? With the recognition that the historical timing of "globalization" is fluid, it is important to note that in this book I will focus on globalizations' intensification, beginning with the rise of neoliberalism with Reaganism and Thatcherism in the early 1980s.

Globalizations "from above" and those "from below" are not geographic, nor do the terms point to hierarchy of origin in relation to another point (e.g., a powerful entity such as UNESCO versus governance within local communities). This is indicated by utilizing the plural "globalizations," which indicates that processes of globalization can be either empowering or disempowering (e.g., globalizer/globalized and from below/from above) and thus demand rigorous, contextual analysis to better understand who/what[3] they affect negatively or benefit (Torres, 2009). In understanding how current global processes (socio-environmentally) affect local societies, critical historical analysis is essential (Misiaszek, 2020, 2018b).

I have been questioned on the from below/above arguments by readers of my last book on ecopedagogy and global citizenship education (GCE) (Misiaszek), which I want to clear up a bit. Similar to the non-geographic Global South and Global North that de Sousa Santos (2018) uses to differentiate epistemologies of the South from epistemologies of the North, as was discussed previously, globalizations from below and from above are countering or coinciding with oppressions/domination, respectively. Globalizations from below focus on globalizations effects upon the local but also upon the widened planetary sphere within ecopedagogy, in the same ways "the South" (see de Sousa Santos (2018)) does not refer to geography but rather to the oppressed, the exploited, and the subaltern. "Below" also refers to the locations of oppression by globalization from above, and I would add that "below" also includes the non-anthropocentric sphere that is dominated, leading to local to falsely justify actions leading to planetary unsustainability. The wording of "the South" and "the North" will be further described later in this book by utilizing concepts by de Sousa Santos and others.

"From below" constructions of globalizations have been discussed in various ways, for example, with the following excellent definitions.

> "Globalization from below," or "anti-globalization," which we see as a misnomer because various groups and movements whose aim is to challenge neoliberal versions of globalization are not opposed to increased international integration in general. Globalization from below is largely manifested in individuals, institutions, and social movements that are actively opposed to what is perceived

> as corporate globalism. For these individuals and groups, "no globalization without representation" is the motto. (Torres and Rhoads, 2006)

> As Appadurai [(1996, 2001)] suggests, both a deparochialisation of the research ethic and attempts to give real substance, beyond rhetorical gestures, to the idea of "globalization from below." This demands an understanding of the differentiated impacts of neo-liberal globalisation on the first world and the third, an effort to take seriously the needs and interests of the marginalised nations within the narrative of global capitalism, and an dialogic engagement with educational policy researchers in global networks in a manner that is both serious and reciprocal. (Rizvi, 2006)

Centering the framings of globalization from below on reading globalizations' effects of unsustainable environmental violence through these two quotes, ecopedagogues must teach to read how globalizations affect socio-environmental violence, dominance, and (un)sustainability to determine possible ways to utilize the processes to counter those from above.

That globalizations from above value certain constructs and devalue all others is essential for their continued existence. By far the most prominent example of globalization from above is Westernization; however, as stated in the previous below/above globalizations framings, it can also be from below, depending on the processes at hand. In other words, just because globalization emerges from a local community it does not mean that it is countering oppressions/dominance. For example, if a local population states that it is their culture to not educate females, the process would be termed globalization from above, although locally contextual. However, it is essential to problematize through reading who proclaims this as "cultural" and who benefits from it—with the most likely answer being patriarchy within the local society. This also calls upon the need to read through the sociohistorical aspects of "what" is being "globalized," such as problematizing with de Sousa Santos' (2018) arguments of oppressions emergent from different sides of the "abyssal line," corresponding to questioning the epistemological entrenchment of histories of colonialism, patriarchy, and capitalism.

In understanding globalization and playing off of Anthony Gidden's (1990) famous quote,[4] ecopedagogues teach to read with the goal of understanding how it affects local context, as in the ecopedagogical call for deepening socio-environmental understandings from widened global influences. Such deepened understandings of globalizations are first needed to know how the processes affect people and Nature, to also widen understandings to the global (i.e., the world) and beyond Earth overall. This incorporates many aspects, including what we value, both individually and from the world holistically, and how we comprehend "development," "sustainability," and "livelihood," to name a few.

For example, to teach deepened and widened understandings of sustainable development and globalization, ecopedagogues must problem-pose how, as a society, we determine what to sustain and at what baseline of sustainability. Are these determined through local and/or global lenses, as well as are they determined anthropocentrically or planetarily? I argue that making this determination through global lenses creates a deficit-framed determination of sustainability because global demands on a local society are almost always impossible and the structurally "exportation" of socio-environmental violence (from central to peripheral States, from the globalizers to the globalized) creates local unsustainability. As will be discussed later, when reading, teaching, and researching "sustainable development," it is essential to critically deconstruct "whose" development/progress, "whose" sustainability, and according to what baseline we focus sustaining at. And, considering the continuum of global to local spheres, at what sphere(s) is sustainability considered and how does globalizations affect this analysis of sustainability?

There is also the difficult question of what/who the agents of globalizations are, who possibly has metaphorically outspoken voices, as educational comparatist Nelly Stromquist (2002b) has argued that globalization is "not a decentered phenomenon but rather has definite points of origin." What/Who are the agents of globalization, or possibly more the classifications of them and/or their degrees of influence, are argued between scholars; however, it is this complexity and ambiguity that sustains much of globalizations from above. Ecopedagogical work is to unveil them through reflectivity of the limitless complexities. It is important to note that when we are discussing globalizations it is well beyond governmental entities and governmental-based international organizations, such as the United Nations.

Stromquist (2002b) has given five agents of globalizations as the following: the States, transnational corporations (TNCs), mass media, nongovernmental organizations (NGOs), and global financial institutions (e.g., the World Bank). There are numerous difficult and contextual questions regarding what decision-making influences, if any, do the global masses have upon decision-making from above that affect them (in)directly. Such decision-making at long distances which affects the local with little or no democratic input, coincides with Gidden's (1990) famous definition. Globalizations widen the questions on environmental decision-making, such as what voices are "listened to" within one's own national and subnational governments? Relating back to the questions in the previous paragraph—with increased distancing, what voice does someone have with a State that they are not an official citizen of, for example, what voice in environmental decision-making does someone outside of the United States have on the environmental violence committed by the United States within the global/planetary sphere?

Inside and outside States' (e.g., nation-states, city-states, etc.) governments, ecopedagogical reading of global governance from and between TNCs, NGOs, mass media, and global finance entities as having local impacts on the non-anthropocentric sphere is essential, but limitlessly complex and difficult. An example is reading to determine what the influences of TNCs upon environmental violence and dominance are. Many describe TNCs' influences as the most powerful due to their vast economic means and international reach, especially with increased blurriness of specific governmental control within an increasingly globalized world. For NGOs, a key question is who are the decision-makers—is it bottom-up decision-making including those who are directly impacted or are they boards with a few powerful individuals as members? For example, what are the influences of education globally from the Melinda and Bill Gates Foundation and, the follow-up question of, who are the Foundation's decision-makers?

For mass media, what are the hidden controls of the distribution of knowledge and (de)legitimization of knowledges and ways of knowing, especially when discussing the internet? Teaching ecopedagogical literacy is Freirean reading outside of formal and nonformal education spaces including deconstructing public pedagogies (i.e., informal education). The media provides an education about how to interpret reality, but what are the influences upon media which develop such realities? How are acts of environmental violence, social injustices, and connections between them portrayed by the media, as well as the influencing forces that systematically justify them, and the connections between them? Critical media culture theorizing allows for systematic uncovering of a "truth" or reality outside of hegemonic hidden curricula that form a person's understanding of the world-Earth (i.e., thematic universes, as Freire would name them in dialectic teaching (see Chapter 3)). In other words, how do you teach students individual reading through ecopedagogical literacies in tangible spaces, where "teachers" are also "students" and vice versa, in research fields and in methodological spaces of analysis, and within "virtual" spaces such as within social media and online content development (see Kahn and Kellner (2006)). Here, ecopedagogical spaces are ones that are not passive but active; however, teaching ecopedagogical literacies to "read" media that are passive mediums requires active engagement, with "media" indicated here as vast as possible (including this book) (Kahn, 2010). The media is a powerful public tool to develop a sense that environmental devastation is necessary; and often hides connections between environmental and social problems. What control and associated global power does Google have in the search for information? What if something comes up in the first ten listed in a search result compared to the 160,080th to the 160,090th?

Lastly, who controls the global world financial systems, including the World Bank which is the largest educational research entity in the world? There is a vast amount of questions on the World Bank on educational decision-making, including the problematizing of the fact that their research is largely not blindly reviewed, but rather carefully selected by reviewers if found to be within "agreeable" politics (Arnove et al., 2013; Klees, 2008b, 2008a). Theories to deconstruct globalizations' influences upon the world and Earth help guide informed socio-environmental praxis, to better understand the barriers of democratic influence as Freirean limit situations to determine how to increase possibilities of local-to-global democratic participation.

On the topic of the complex ways in which "development" is both constructed and carried out, the local-to-global to planetary complexities of who is benefiting and who is suffering is endless, especially with the understanding that development is endlessly contextual. I term it "endless" because we individually and societally need and want different aspects that conflict and coincide with others. We return to the question of the level or framing of "development" that we need to sustain. Teaching through difficult questions is almost a given here because the question is, what if lower 90 percent of the population, in terms of economics, had the lifestyle of the top 10 percent? In this question, there is the question of needing to lower one's lifestyle as largely self-defined within current framings of development, in populations of the top.

Ecopedagogically reading development and sustainability includes the need to approach how globalizations affect how the world defines them and the complexities of increased distancing of populations being negatively affected for others to benefit. Within an increasingly interconnected world we need to understand these aspects both as contextually without our own identities/personhood and as world-Earth balance of "our" and "their" development and livelihood. This is exemplified by Carlos Alberto Torres' argument that citizenship education should focus more on personhood with individuals having multiple identities, for reading and teaching to understand and respect persons' diverse identities in an increasingly globalized world (Torres, 2002).

This aligns with Raymond Morrow's (2006) argument that diversity and autonomy of personhood with individualized thought is also an enemy of neoliberalism, especially when social justice is the goal within education, due to its consistent, and very much necessary, countering of systematic normalization of neoliberalism. Along with most critical pedagogues, I argue that neoliberalism only works when teaching resists questioning its ideologies, within learning spaces and suppressing questioning outside of learning spaces, to instill "no alternatives" to neoliberalism and neoliberal Development ("D" uppercasing will be explained in next subsection).

Ecopedagogical literacy includes reading the non-anthropocentric sphere to critically problem-pose what is necessary for planetary sustainability as globalizations from above, especially neoliberal globalization, largely ignore world sustainability with the non-anthropocentric sphere and, without needing to be stated, planetary sustainability. The last few paragraphs have discussed opposing aspects of decision-making of socio-environmental justice within globalizations from below. However, it is essential to remember that ecopedagogical reading also includes how globalizations from below/above within the world affect the rest of Earth. If our actions are sustainable or unsustainable for Earth overall, the way in which globalizations affect decision-making is essential to read. However, it is also important to read that social interactions of the world have no consequences on the effects of environmental violence for the non-anthropocentric sphere. For example, if it is democratically agreed upon to cut down almost all the trees within a forest, desertification and/or increased mudslides (among other environmental disasters) would still occur, even though the actions themselves were locally determined through democratic means and socially just in some, or even many, regards. Globalizations likely place more pressure for lumber and fuel needs and wants which complicates the already difficult ecopedagogical reading of world-Earth (im)balance, (non)violence, and dominance or the lack of it. But putting these issues temporarily aside, the local-to-global effects of the actions within the non-anthropocentric sphere and overall unsustainability is "real" regardless. The ecopedagogical utopic goal is that critical reading with deepened and widened understandings will lead to better decision-making for the well-being of the non-anthropocentric sphere, resulting in a thickened and widened democracy through more diverse voices and epistemologies. I would argue that within ecopedagogical work, this should be the goal of globalizations—a defining factor of globalizations from below.

Widened literacy, including widened ecopedagogical literacy, within an increasingly globalized world, along with deepened contextual readings is difficult, to say the least, but essential. Connected with these and other environmental issues of globalizations, I will later discuss possibilities but also challenges of widening ecopedagogical literacy through reinventing already known concepts—such as citizenship, which has been a sociohistorically contested terrain of inclusion and exclusion. These discussions will include various theoretical framings including global citizenship and its education (GC/E) and planetary citizenship and its education (PC/E).

Asking how such reading is both more easy and difficult in an increasingly globalized world leads to endless questions and conundrums. For example, how can globalizations from below help us better ecopedagogically read the world-Earth as calling for reinventions of globalizations? Increased mobility, access information, and communications are unquestionable tenets of most globalizations, among others, but there are also limitless processes of manipulation, further division, and decrease in meaningful critical dialogue—among numerous other negative aspects. There is no single way or path toward such multiple, intersectional spheres needing to be critically read, but I write this section, and this book, to hopefully provide some key tenets needed within ecopedagogical work. One way to describe the tenets of reading of and through globalizations is global meaning making, as described below by Tierney (2018b), in which reading global and planetary inclusiveness (i.e., "widening" within this book's terminology) is seen as connected to better understanding one another (i.e., "deepening"):

> Global meaning making entails forms of transaction that occur in support of a reckoning of oneself and one's cultural ways of knowing as one journeys across borders with others and for the interests of all. It represents a mix of participatory literacy promoting approaches that are cooperative, collaborative, and contrastive but respectful and reciprocal. It befits a planet that is ecumenical and emancipatory.

My discussions on the deepening and widening aspects of citizenship, development, and balance/sustainability with the rest of Nature throughout this book, I argue, coincides very closely with Tierney's constructs of global meaning making, although with differing terminologies (e.g., "deepening," "widening"), with the same overall goals.

Returning to the topic of "development," problematizing what is "development" and who it is for, needs to be through diverse epistemologies. There are many reasons for this, including to better understand both the commonalities and the differences between how diverse populations view socio-environmentally "good" development as essential with environmental issues that are almost always never contained locally and are often global in effect. As the world "develops" with one of education's key goals as "development," *how* it develops is governed by education in defining its endpoints will all these aspects as largely steering toward or away from educational violence. In other words, how do we locally-to-globally-to-planetarily read development to determine what violence is "needed" and is "acceptable," with the reasoning that environmental violence is a continuum. Environmental violence is almost never contained locally, but rather spreads well beyond geopolitical borders with globalizations facilitating, often manipulating, the mechanisms of greater global distances—which I have previously defined as "distancing" (Misiaszek, 2018b). For example, what environmental violence has the United States exported to other countries in terms of mining for natural resources and the production of products? Who is affected by such environmental violence so that those residing in the United States are not affected by such violence and that monetary profit can be increased?

Coinciding with the work of several postcolonial scholars, there are severe limitations of educational comparisons only between nation-state units in an increasingly borderless, globalized world. Such comparativist modeling makes it easier to ignore (neo)coloniality. The work of Roger Dale (2005), and other comparative educational scholars, has expressed the need for comparing educational systems within structures of local-to-global governance rather than within and between traditional geopolitical borders—as though globalization does not exist. Within environmental problems, as argued before, geopolitical borders are largely meaningless in many regards.

Ecopedagogical reading of globalizations includes problematizing (un) democratic structures of governance in which giving, amplifying, and silencing voices of those affected by environmental violence are essential dynamics to determine, as is ecopedagogical work for praxis on how globalizations from below can allow for widened democratic participation with amplified voices from those often silenced. Key problem-posing questions include the following: Who has a voice(s) among those who take on such violence upon themselves, from local-to-global spheres? What are the global structures that allow for exporting such violence, within the concepts of distancing? If the voices aren't from those who suffer from the violence and are not "heard" and acted upon within democratic structures, then the globalizations are, without a doubt, globalizations from above. Although the necessary social structural transformations are very difficult from the voiceless-ness of global governance structures, what ecopedagogues must teach toward are goals to disrupt framings of development, progress, globalizations, and modernities that lead to socio-environmental violence, dominance, and unsustainability (Misiaszek, forthcoming, 2020, Misiaszek, 2020, 2018b).

The complexities of making meaning of globalizations within and between local contexts are difficult to fully describe, so I am going to utilize Tierney's global meaning making for describing them. As ecopedagogues are critical global educators and researchers that seek to deepen and widen understandings of the world, they must be global meaning makers who "straddl[e] multiple locales in different countries, including spaces where racism, classism, and ethnic and various other forms of discrimination may be deep-seated, almost intractable, and perhaps perpetuated unless challenged" (Tierney, 2018b). To express the complexities of global-local interconnectivities I have used some of Tierney's descriptions, who has utilized the work of several other critical scholars, of global meaning making which is essential within ecopedagogical work.

> [global meaning making] need[s] to be diverse rather than standardized, ever changing and continually adapted, especially by the bottom-up and sideways flows of people, ideas, and acts informed by cultural appreciations and ethical tenets (Abdi, 2015; Campano et al., 2010; King, 2017; Tierney, 2017a, 2017b, 2017c, 2018b)
>
> . . . transliteracy framework by Stornaiuolo, Smith, and Phillips (2017) that explores dimensions such as emergence, uptake, resonance, and scale as a way of capturing "different kinds of relations among people and things—whether in horizontal, vertical, rhizomatic, or other relationships—and highlight(ing) people's literacy practices within and across systems that (re)produce, exacerbate, and/or challenge social inequities" (p. 84) (Tierney, 2018b)

Tierney and the scholars he cites do not use the same terminology that I use in my work (e.g., transliteracy), including this book, but the essence of the themes expressed as global meaning making very much coincides with many of my descriptions and arguments for ecopedagogical work.

I want to end this section by reiterating that the ecopedagogical goals of deepened and widened understandings of the world-Earth through diverse voices and epistemologies, along with amplifying the sociohistorically voiceless to widen and thicken democracy, must be defining factors as we speak of globalizations from below. As such, ecopedagogies are inherent enemies of globalization from above, especially neoliberal globalization. Teaching to read the politics of environmental knowledges will be what the next section will focus upon, with particular attention on how globalizations from above is the thick residue from colonialism (i.e., coloniality) and thus can be rightfully termed as "neocoloniality."

Education that systematically manipulates realities of the globalized largely sustains and intensifies socio-environmental violence, dominance, and planetary unsustainability. Ecopedagogical work deconstructs such epistemological manipulation of socio-environmental knowledges. This is very strongly stated by Freire (2000) that "the invaders desire to know how those they have invaded apprehend reality—but only so they can dominate the latter more effectively."[5] For example, epistemological manipulation toward domination through global *D*evelopment causes de-*d*evelopment. ("D" casing will be explained in the next subsection.) As will be discussed in the next section, gathering knowledge from the colonialized/globalized was almost never for better understanding in building ecologies of knowledges, but rather a tool of dominating manipulation. The (neo)colonizer inserted its *S*ciences and *K*nowledges for oppressive control, including the manipulation of the (neo)colonialized *s*ciences, *k*nowledges, and *t*echnologies.

2.2.1 Development and globalization: Introducing *d*evelopment and *D*evelopment

There are many globalizations' influences, including how globalizations from above systematically narrow frameworks to a single, dominant form of *D*evelopment (indicated with the uppercase "*D*," which is also *underlined and italicized* throughout my work and in this book). Opposite to and inherently opposing *D*evelopment is lowercased *d*evelopment. Ecopedagogical reading of the sociohistorical aspects of *d*/*D*evelopment is essential to understand how the hidden past oppressive structures continue currently and into the future, for example, coloniality's exploitation and raping of natural resource follow globalization from above, and is suitably named neocoloniality.

Globalizations, similar to coloniality with specific aspects of globalization from above as equated to neocoloniality, need to be theorized *through* critical theories as they are entrenched in almost all aspects, all spheres of the world. Toward the goal of ecopedagogical deconstructing of development, reading through trans-theorizing is essential to teach through and between theoretical frameworks to understand power dynamics resulting in socio-environmental violence and planetary unsustainability/dominance by valuing one's population's(s') "development" over others' "development." Later and throughout this book, the use of theoretical lenses (e.g., critical race theories, feminism, coloniality (e.g., post-, de-, neo-, anti-)) within ecopedagogical work will be discussed more to determine how socio-environmental suffering can be understood through multiple perspectives.

The following is a quote from my previous book on the (neo)coloniality of globalizations from above, emphasizing the need for a critical sociohistorical reading of globalization.

> What is necessary in analyzing globalization within a "postcolonial" world is how processes of globalization can counter colonial-structured education oppressions rather than often sustaining such oppressions with neocolonial aspects (Abdi, 2008). (Misiaszek, 2018b)

It is the continuing oppressive past that is systematically taught through hidden curricula that promotes the false truths that colonialism is in the past without current relevance (Dussel, 2012; Grosfoguel, 2008); however, its thick residue remains today in the form of global exploitation of the world's "weak States" (i.e., the globalized, "peripheral States") by the "strong States" (i.e., the globalizers, "central States"). I have emphasized this in the following quote.

> Without problematizing hegemonic Development models, environmental pedagogies become oppressive and reproduces neocolonial structures (Misiaszek, 2018b). . . . How globalizations center or decenter distant oppressions, many of which can be juxtaposition to colonial oppressions and labeled as neocolonialism, is essential to problem-pose in teaching "development"—does the "progress" legitimize inequalities of world systems or helps to end local socio-environmental oppressions (de Oliveira Andreotti, 2011a; Arnove, 1980; Kellner, 2002; Stromquist, 2002b; Torres, 2009)? In questioning the socio-environmental effects due to globalization, the question is the same for globalizations' contested terrain as the defined development centering or decentering Central States' global hegemony (e.g., the Global North). If education should lead to development and this "progress" is related to modernity, problem-posing globalizations is essential to determine if the processes construct a single linear path, such as toward Westernization, or can they decenter a single pathway to modernity defined by Westernization (Arnove, 2007; Cudworth, 2003). Aligned with this question on globalizations, which processes focus only Development as an endpoint and which processes provide possibilities for development. In ecopedagogies a singular, linear path of development within a socio-environmentally diverse planet must be problem-posed in ecopedagogies, as well as the oppressions of the endpoint, such as Westernization, which can be empowering but also oppressive. To be clear, this does not mean all of Westernization is negative, but it certainly does mean that

> not everything in Westernization is beneficial—and here ecopedagogies are needed in order to decenter an ideology that Westernization is the expectant outcome of progress and modernity in an increasingly globalized world. (Misiaszek, 2018b)

An essential ecopedagogical problem-posing question is whether defined "development" centers or decenters the global hegemony of central States (e.g., the Global North). Much of my previous research has found ecopedagogues stressing that a singular, linear path of development within a socio-environmentally diverse planet must be problem-posed in environmental pedagogies, as well as the oppressions of development endpoint(s) such as Westernization, which can be empowering but also oppressive within models of neocoloniality (Misiaszek, 2020, 2018b). Reproducing environmental pedagogies grounded in neocolonial ideals helps to sustain coloniality in all its forms, as well as its veil*ness*.

2.3 Economics: Countering neoliberalism

> [Neoliberal globalization:] It's a question of jumping on the train in the middle of the journey without discussing the conditions, the cultures, or the forms of production of the countries that are being swept along. And there is no talk about the distance that separates the "rights" of the strong and their power to enjoy them from the fragility of the weak in their attempts to exercise their rights. Meanwhile, responsibilities and duties are leveled-equal for all. If globalization means the abolition of the frontiers and the opening without restriction to free enterprise, those who cannot compete simply disappear. (Freire, 1998b)

The normalization of neoliberalism that deems all other economic structuring as "abnormal" calls for ecopedagogues reading that counters such ideologies through problematizing the inherent aspects of neoliberalism that devalues world-Earth sustainability and socio-environmental justice/peace for economic profit, as well as associated power. The "jumping on the train," as described by Freire above, is for economic profit not of those passengers getting on the train abruptly but rather for a few others to sustain/intensify their hegemony, while falsely telling the new passengers they will benefit and to jump on is "the only way." Radical transformation that ends local-to-global systems of neoliberalism is a central ecopedagogical goal.

Countering the normality of neoliberalism is an essential aspect of ecopedagogy with both students and teacher(s) needing to critically read and reread what is progress and what is livelihood through local and global lenses that can develop into counter-stories of neoliberal framings of both (Misiaszek, 2020, Misiaszek, 2018b). Ecopedagogical reading of "development," "sustainability," "livelihood," and counter-stories of those most suffering from framings that oppress them will be introduced in the next section. Determining who suffers and who "most" suffers from instilled ideologies taught must be problematized—aligning with the arguments throughout this book. Teaching to question the ideological and epistemological foundations of societies questions the very systems of unjust power structures and, within the ecopedagogical aspects, the rationality of unsustainable environmental violence. Without countering such aspects in transformative, radical praxis, such unsustainability will remain. Environmental pedagogues are often the culprits in sustaining and/or intensifying socio-environmental violence consciously or unconsciously, by not being critical or transformative within their teaching of normative ideologies (Misiaszek, 2018b). Although such problematizing is essential in all education, school classrooms are often absent of this, and frequently suppress such dialogue; as Freire (1998b, 1998c) and Apple (2004, 2014) have argued schooling curricula regularly encourages/forces false neutrality in classrooms. This is more evident in dialogue that counters "official" knowledges.

The following two quotes discuss the need for radically transforming economics by recognizing other systems in which environmental justice and humanization are foundational, as well as valuing of the rest of Earth (i.e., planetarization).

> We need to recognize that they both promote "other-than-capitalist" ways of being and strive to redefine life. The life-projects that they seek to defend, rediscover or invent anew share similar ontological orientation and all are expressions of movement against alienating and objectifying effects of capitalism. I would like to provocatively offer that EJ [Environmental Justice] and degrowth are united by a common "striving," for autonomy, connections, and well-being, a striving laden with potential for alternate world-making. (Singh, 2019)

> From an environmental economics perspective, environmental degradation can continue so long as overall stocks of capital are increasing and mobilized to technically "solve" environmental problems created by economic activity (Spash, 2013). In this manner, "Nature" is viewed as outside of humanity, abstracted and reified from its social context, and made measurable and fungible as a commodity in the marketplace. In such a vision, humankind is effectively alienated from the nonhuman world which serves as its original reference point for all creativity (Becker et al., 2005). Instead, Natures conceived as non-human living beings, abiotic matter and phenomena, and the human being itself as labour, are treated

> in the ensemble as objectified inputs for economic production. (Swyngedouw, 1996; Kolinjivadi, 2019)

Capitalism has a range of effects on environmental pedagogies, including normalizing the commodification of Nature with neoliberal goals at its word, thus (in)directly opposing world-Earth sustainability. The objectifying of humans, their labor, and the rest of Earth that is foundational to capitalism, and its more extreme framings of neoliberalism, can be read throughout these two quotes. Ecopedagogical work focuses on critically reading the factors of current economics systems within an increasingly globalized world. Ecopedagogical reading asks what economics "should be" through multiple framings, including ecological/environmental economics which Kolinjivadi (2019) eloquently described in the second quote. Kolinjivadi (2019) explains the dualism of arguing for the sake of, with accompanying economics of, "society" and "Nature"—a topic that will later be discussed at length within the aspects of economics, development, and world-Earth (i.e., world-Earth de-distancing) throughout the book, including the next section on reading *d*/*D*evelopment.

Ecopedagogical work unveils hidden economic framings and ideologies that form the landscapes for socio-environmental oppressions, dominance, and unsustainability, including their entrenchment by banking-education structures. This entrenchment has been discussed by many critical scholars, including Moacir Gadotti and Torres (2009) as they exemplified Brazil's need for alternative economic systems taught through ecopedagogies to counter past and current (neo)colonial banking schooling (if schooling was provided at all) that have not prepared the general public for democracy or true *d*evelopment.

As Freire (1998b) discussed neoliberalism as stepping onto a train without knowing its destination in the quote previously given, the faith of neoliberalism in the market, especially the global market, is without any concern for humanization, or planetarization within ecopedagogical work. De Sousa Santos (2018) wrote that "there is resistance [to neoliberalism], but it is less and less credible as a bearer of a realistic alternative." The quote below indicates neoliberal measuring "value" through economics in which the justice, sustainability, and equality is not considered within the measurement, at best a far second in priority and often much farther behind than second.

> Mechanists and humanists alike recognize the power of today's globalized economy. However, while for the former there is nothing to be done about this untouchable power, for the latter, it is not only possible but also necessary to fight against the robust power of the powerful, which globalization has intensified, as

> it has the weakness of the fragile. If economic structures indeed dominate me in such a masterful manner as to shape my thinking, to make me a docile object of their power, how can I explain political struggle, and above all, how can struggle be undertaken and in the name of what? . . . To me, it should be undertaken in the name of ethics, obviously not the ethics of markets but rather the universal ethics of human beings—in the name of the needed transformation of society that should result in overcoming dehumanizing injustice. (Freire, 2004)

As I discussed in previous publications (Misiaszek, 2020, 2018b), critical pedagogies and environmental pedagogies are often criticized for not valuing the teaching/reading of economics, but I argue for the opposite, with economics being a central priority within the pedagogies, especially ecopedagogies.

De Sousa Santos called for alternate models of economics that is not based on epistemologies of the North grounding capitalism, as well as histories of colonialization and patriarchy, but rather opposing them (i.e., emergent from epistemologies of the South).

> Social and solidarity economy can express the various forms of grassroots, peasant, indigenous, and communal economy and the kinds of property associated with them, different among themselves but, as a rule, anticapitalist and anticolonialist (and often also antipatriarchal), based on principles of reciprocity and relationality at the antipodes of capitalist and colonialist logics. (de Sousa Santos, 2018)

Ecopedagogical praxis, in part, should be toward this end. Specifically, ecopedagogical work is to problem-pose systems of economics to seek possibilities and current aspects of socio-environmental justice and sustainability—for example, problematizing whether local-to-global structures of monetary exchanges lie within or outside framings of economic justice, labor justice, and "sustainable development." Do our economic structures value leading toward or away from, or absolutely ignoring, world-Earth sustainability? What epistemologies and ideologies are legitimized to both "know" about, and emerge from, economic systems—a question of widening or narrowing one's Self and others' reflectivity?

In my previous publications (2020, 2015, 2014a, 2016a) I have also discussed how even within economic framings, socio-environmental violence does not make sense, especially if you lengthen the observed timeframe or widen the populations affected. For example, global-warming denial is a short-term financial gain for some populations and individuals, but not for others. But over a longer time period all the world's populations will be affected. Initially,

I considered writing that only "a very few" benefit from global warming, which is true within the largest of profit gains, coinciding neoliberal tenets. However, there is a large percentage of the world's population that benefits from relatively cheap prices of fuel and natural resources, including myself. Currently, I am writing this book and various messages on Apple products which are relatively expensive, but extremely inexpensive when considering the natural resources within them that will be increasingly costly for many in economic terms, including, but not limited to, environmental violence and labor injustices. I benefit from such socio-environmental acts of paying less on the back of those who are often the most vulnerable, and also those who will be affected more and more from environmental devastation, which would include me over time.

Later in the book I will discuss in depth the need to critically deconstruct the inequalities of "benefiting" and "suffering" from the mining and use of natural resources, especially when comparing effects of sociohistorically oppressed populations along the lines of gender, race, sexual orientation, ethnicity, global positioning, and other ways we artificially categorize the world and accompanying social roles. A critical economic question, beyond justice models of economics, is, what will be the monetary costs of rising ocean levels, from global warming? Taking, for example, New York City, which is predicted to be largely underwater once the polar ice caps are melted, what are the financial costs of such an event?[6] With a bit of dark humor, I sometimes joke during presentations that the only beneficial aspect of Trump's global-warming denial is that Trump Tower will also eventually be underwater. More devastatingly in the near future, islands, and populations living on them, will lose entire societies when they disappear under water. What are the economic costs of such environmental events, which could be even problem-posed within neoliberal economics of benefiting a very few elite, and whose long-term effects will also be negative to neoliberalism's benefactors? Of course, the effects of those less locally to globally powerful will be much more devastating.

It is important to remember that ecopedagogical work on economics must be extended beyond the world to the non-anthropocentric sphere, in how "we" rationalize environmental violence for economic reasons. Rationalization within neoliberalism is only through satisfying its tenets of profit (as described previously) for the Self and/or the Self's private sphere. Dirk Postma's (2006) argument that neoliberalism's focus is economic profit within the private sphere to devalue all public spheres can be extended to the largest of spheres—the planetary sphere (i.e., Earth). With non-valuing to a point of nonrecognition of local-to-global sphere issues beyond economic and power gain of one's own private sphere, the non-anthropocentric sphere is simply not considered. Aligned with some of Freire's first arguments in *Pedagogy of the Oppressed* (2000) in problematizing why "success" is viewed by the number of people beneath the Self, individualized economic "success" is measured on increasing inequality with "me" to distant "others" in relation to economics and (in)direct associated power that emerges from such inequalities. Expanding this beyond the world, the non-anthropocentric sphere is viewed only to be utilized for economic/power gains to increase inequalities. World-Earth sustainability is only considered within neoliberal framings if it profits the private sphere, and simply only that.

With economics so ingrained within our societies and environmental actions, this problematizing, in part, necessitates radical transformation of environmental pedagogies. Ecopedagogies teach for possibilities of transformation of all, as opposed to shallow environmental pedagogies that solutions can only emerge within current economic structures and ideologies within those structures—among other societal structures discussed previously. Returning to the false critique of environmental pedagogies ignoring economics at this chapter's beginning, ecopedagogies do not teach economic ideologies to reproduce socio-environmental violence but rather teach to read them to transform hegemonic economic ideologies to end the unsustainable violence emergent from them.

Economics entrenched with global neoliberalism both strengthens this need and almost limitlessly complicates teaching with well-hidden global politics, as well as obfuscating praxis with distancing of democracy within global governance, as argued previously. Globalizations from below can have many opposite, countering effects. De Sousa Santo's (2018) quote eloquently expresses the sociohistorical oppressions from colonialism constructing neocolonialism (without stating the word directly) as it is connected to governing violence.

> The crisis of governability that, in one way or another, is present in contemporary societies is the result of a historical condition intrinsically linked to the current phase of global capitalism (neoliberalism) in which knowledge as regulation is poised to free itself from its counterpoint (knowledge as emancipation) and, as a result, to produce a kind of order structured by the duality between appropriation and violence, the duality characteristic of colonial regulation . . . linked to guaranteeing a type of order possible only through appropriation and violence.

Ending reproductive education, including reproducing environmental pedagogies that sustain and intensify violence by appropriating neoliberalism, without offering alternatives, is an important tenet of ecopedagogies. As will be discussed later, ecopedagogical work includes the problematizing of environmental pedagogies (and education on the environment with the unfortunate role of sustaining socio-environmental violence through neoliberal justification, as well as fatalism of non-transformability of the world systems, including economic systems (Misiaszek, 2018b, 2011)). This coincides with Freire's (2004, 2000, 1998b, 1992) writings on fatalism ingrained within schooling systems, especially within banking-education models, with unchallenged, continued histories of oppressions.

I end this section with de Sousa Santos' quote (2018) expressing unabating condemnation for capitalism, followed by some discussions on how his and others' arguments will be incorporated into other previous and upcoming topics of this book.

> Not questioning the possibility of capitalism reproducing itself indefinitely, would affect its drive for concentration of wealth. I have in mind European-style social democracy. Having gotten rid of these two threats, global capitalism seems to be thriving in spite of (or because of) being permanently in crisis. A permanent crisis is a new type of crisis. Instead of demanding to be explained and calling for its overthrow, it explains everything and justifies the current state of affairs as the only possible one, even if it involves the imposition of the most grotesque and unjust forms of human suffering that were supposed to have been thrown into the dustbin of history by the progress of civilization.

The reproduction of capitalist ideologies leading to the "most grotesque and unjust forms of human suffering," as it is embedded within governance and democracy, should not be "thrown into the dustbin of history" without critically understanding how economics system lead to such suffering. De Sousa Santos (2018, 2016, 2007b) develops many arguments throughout his writings on how dominant epistemologies (i.e., epistemologies of the North) ground such ideological reproduction through academic disciplines, despite the fact that many academics and public intellectuals focus on ending oppressions. In Chapter 5 I will further discuss de Sousa Santos' (2018) call to determine the "absences" within disciplines for them to be reinvented through epistemologies of the South (i.e., "emergences of disciplines," coinciding with emergences of sociologies) as necessary toward terminating colonial, gender, and capitalist oppressions within academia and its tentacles. The epistemological foundations of many societal structures, including economic systems, have emerged, as de Sousa Santos (2018) argues, from the oppressive structures of capitalism, colonialism, and patriarchy, which does not allow for ways of knowing to end oppressions because the "knowing" is grounded in the oppressions. While it is true that various critical scholars argue against de Sousa Santos' rejection on many foundations of sociology, especially when done so broadly, his argument nonetheless provides insight into the foundations on how we understand societies, including ourselves and "others" within them. Teaching through such epistemological deconstruction is essential and rarely done within formal education, especially below tertiary level, because schooling is often, to varying degrees, an ideological apparatus of the State in which ideologies are to be taught not questioned. Regarding economics, the problematizing initially involves the epistemological foundations of economic reasoning and whether the economic actions are coinciding or contrasting with the reasoning, with what is stated compared to what is actually happening.

2.4 Development and citizenships: An introduction

I begin this section by returning to the aspects of balance between continuums of environmental violence that humans commit upon the rest of Nature to survive and fulfill their diverse notions of "livelihood" and "development," among other human-centered notions of needs and wants. In various other works on structuring ecopedagogy, I have centered the two following foundational aspects. First is that all environmentally violent acts are done to benefit someone, with the "someone" as a range from an individual to a human population that could be as large as the world in its entirety. With this last phrase written, I cannot think of a case in which an environmentally violent act benefits all of the human population. This is a key dilemma with the world sphere—what individuals/population benefit and which ones suffer? What are the politics of environmental actions is the overlaying question to be problematized within ecopedagogical spaces of learning, reading, and research? Examples that give some sense of the continuum to environmentally violent acts are: cutting down a tree to burn the wood for warmth, the fuel used to make electricity to write this book, and mountain top removal (MTR) for coal. Ecopedagogical work deconstructs "who" is benefiting, how, and to what degree of (in)equality with those who are suffering, as all these are acts of environmental violence that benefit certain individuals/population.

Within the larger planetary sphere, ecopedagogical problematizing includes the effects of such violence beyond our anthropocentric sphere. I always add to this first aspect a disclaimer that our historically limited knowledges and sciences do not always allow us to predict the consequences of our environmentally violent actions. A key question is if our decision-making is with or without hubris. I argue that we often problematically approach environmental decision-making within Illich's (1983) descriptions of "modern man" who believes he (i.e., overall humans) is able to fully control Nature by his intellectual abilities and does not question his own decision-making. Second is the aspect, as previously alluded to, that "our" environmentally violent acts benefit "us," often ignoring how it affects the rest of Earth. The need to critically deconstruct the oppressions of others to benefit is essential within ecopedagogical spaces, but this must be included within the analysis of how Earth is affected. Deconstruction is also needed outside of the anthropocentric sphere—the non-anthropocentric sphere.

I have argued elsewhere that

> if we are to teach to avoid environmental destruction, reconstructing the central goal of education—development—teaching must reconstruct the plural framings of developments through better understanding the tensions between the framings within socio/environmental in/balances. (Misiaszek, 2018b)

I argued with others that two main goals of education are citizenship and development, which are both key aspects of ecopedagogies (as well as all environmental pedagogies) and this book overall. I will discuss in more detail the connections between citizenships and ecopedagogy in Chapter 6, with the re-emphasis here that "citizenship" is to conceptualize how people live together peacefully and productively within groups and that it is best understood as plural (i.e., citizenships) with individuals self-identification as being part of multiple separate and connected identified groupings, from local-to-planetary (see Figure 1.1; the bottom two illustrations *Citizenship spheres (citizenships)* and *Citizenships complexities*, will be more discussed later).[7] With regard to "development," the question is, *whose* "development" is being referred to when we use the term? If development is inherently social as a group's positive progress, the question is, how do we conceptualize inclusion within the group as one's own "development"? In other words, whom do we include in the "whose" as we determine it as "our" development compared to others' development, and what are the separations between them (or, alternatively, problematizing if there should be separations, especially as we delve into "development" within critical global and planetary citizenship models)? The separation between teaching development and citizenship, within such supranational widening of citizenships, is not possible, without the critical understanding on how citizenship is also often a tool for hegemony through division and othering.

Ecopedagogical teaching must widen to include world and Earth "togetherness" constructs, such as global and planetary citizenship, to frame development. Without global and planetary widening of togetherness and connectivity to one another and the rest of Earth, development is incomplete and contextually either positive or negative. As will be discussed later, without such widening, citizenship can be utilized as a tool of global oppression between the citizens and the noncitizen (de Sousa Santos, 2018; Memmi, 1991b), and a tool of justifying unsustainable environmental violence (Misiaszek, 2018b). But what is also important is understanding what is happening within the local spheres as it is affected by the widened spheres (e.g., national, global, and planetary spheres)—that is, rigorous critical analysis on globalizations. Development that is with or without "togetherness" and "connectivity" between the spheres indicates the (a)holistic Nature of how it is framed. Teaching "development" within ecopedagogical spaces calls on the question of "whose" development by problem-posing what spheres of togetherness and connectivity are we posing the question within and between.

For development to be truly "sustainable," its constructs must be widened planetarily in scope and deepened to analyze its effects upon "the local." Without question, teaching between the local and planetary spheres is extremely difficult, but is also exceedingly necessary and should be considered a utopic goal. This goal aligns with Singh's (2019) arguments, presented initially in the Introduction, of NIMBY ("Not in My Backyard") needing to be widened to NIABY ("Not in Anyone's Backyard") and NOPE ("Not on Planet Earth")—the problematizing of "development" must be through these widened frameworks. Such ecopedagogical problematizing is deconstructing *D*evelopment toward reconstructing *d*evelopment, which some scholars (D'Alisa et al., 2015; Latouche, 2004; Singh, 2019) would likely frame as "degrowth"—which I would write as "de-*D*evelopment."

As discussed previously, throughout this book I utilize the lowercased *d*evelopment and uppercased *D*evelopment to indicate, respectively, empowering versus oppressive, holistic versus hegemonic, just versus unjust, sustainable versus unsustainable, and many other opposing framings of who is included within "development" and framings of *d*/*D*evelopment goals. There are no absolute origins or framings differentiating *d*/*D*evelopment, but rather the essence and outcomes of their framings. Returning to concept of "degrowth," this is very much linked to the aspects of development, in which the following quote by Singh helps to explain by framing what degrowth is.

> Degrowth proponents emphasize that degrowth is not "less of the same" or simply a shrinking of material throughput (Asara et al., 2015; Schneider et al., 2010); rather, the emphasis is on "different," different forms of organizing social and economic relations towards "an altogether new, qualitatively different world that will evolve through confrontation with the existing one." (Kallis and March, 2015; Singh, 2019)

Constructs of *d*evelopment that counter growth toward and emergent from *D*evelopment coincide with framings of degrowth as explained above. Ecopedagogical work is aligned with Singh's argument that environmental justice and degrowth movements must work together through dialogue, toward "provid[ing] theoretical tools and imaginaries for rethinking our ways of being in the world and for fostering post-capitalist subjectivities that support socially just and ecologically vibrant communities" (Singh, 2019). Ecopedagogical work inherently counters non-utopic-based environmental pedagogies that teach the world as without alternatives to *D*evelopment, although only used once here in this book, growth which could be written with an uppercase "G"—*G*rowth—aligning with de-*G*rowth.

Globalizations from above and those from below are not origin based or geographic, and polity origins of *d*/*D*evelopment goals and practices are not the ultimate determining factors of the letter-case determination, which is based rather on contextual analysis. This does not mean development's origins are not important but rather that it is not an absolute, binary factor of letter-case determination. It is the critical deconstruction of development through various perspectives, from the local to the global to the planetary, that is essential within ecopedagogical spaces. The difficulty and complexity of this is exuberated by the aspect that *D*evelopment must be portrayed at *d*evelopment, for the least resistance within the public sphere because the true understanding of it as de-*d*evelopment would likely cause resistance to sustaining/intensifying hegemony (Misiaszek, 2020, 2018b). This is further intensified within the current era of post-truthism, a topic that will be discussed at more length in this book's last chapter. The environmental aspects of *D*evelopment must also be deprioritized or vaporized by falsely separating social and environmental violence, and the world and the non-anthropocentric sphere. All of these dynamics are problematized within ecopedagogical spaces to determine *d*evelopment from *D*evelopment and praxis for a world toward *d*evelopment rather than *D*evelopment.

Sustainable development will be the focus of the next section; however, it is impossible to leave out mentioning here that an essential aspect of reading the world-Earth gap is reading how we "read" development. How do we read sustainability within this gap? Within the non-anthropocentric sphere, there is essential critical questioning of what is the value of Earth, which is not the world, with a more popular question of what is the inherent value of Nature, when Nature is categorized as outside of humans. Answering this last question is a difficult one, especially within epistemologies and belief systems that center humans within Earth and often "the universe." The very concept of "value" is one which is to be "placed" onto "Nature," contextually, from humans' populations. As the rest of Nature exists, we as humans are the only ones that reflect upon valuing especially through what Freire (2000) discussed as the human uniqueness of having the ability to reflect upon our histories and the ability of utopic praxis through dreaming, among others. This aspect will be discussed more in the next chapter, with the discussion of world-Earth distancing from the fact that we are the only beings that have the ability to "develop" and "progress" through histories, as the rest of Nature "occurs" and "adapts." This does not devalue the non-anthropocentric Earth, but rather places more responsibility on human actions within the subjective, constructed environments within the static laws of Nature. Humans, as part of Earth, are within the laws of Nature, but returning to Illich (1983) in what he wrote on differences between "primitive man" and "modern man," we have the reflective ability to abide by or ignore them, to our own peril. As will be discussed in detail later, humans "develop" and "progress" as the rest of Earth adapts (Gadotti, 1996; Freire, 1992, 1985; Misiaszek and Torres, 2019; Misiaszek, 2018b); however, human populations' development or de-development is connected to Earth holistically, including how the rest of Earth "adapts." In terms of Illichian (1983) terminology with the realization of his gendered writing, a goal that ecopedagogues teach toward, I would argue, is understanding the world-Earth as the "primitive man" (but degendered) who understood the dangers of challenging Nature as opposed to the modern man who without fear challenges Nature (Misiaszek, 2018b).[8]

It is determining development *with* Earth rather than done *upon* Earth that is essential within ecopedagogical spaces. This includes the changing and often unlearning of ingrained ideologies of our dominance over the rest of Nature, coinciding with what Illich (1983) discussed as primitive and modern man's ideologies of the non-anthropocentric sphere. Teaching toward hegemonic <u>D</u>evelopment follows a false rationality in which contemporary ideologies will never be challenged because they rely on the dominance of the non-anthropocentric sphere to continue. Ecopedagogical reading to problematize what is development at local-to-planetary levels deconstructs what is our relationship to the rest of the world and the larger Earth as we reconstruct framings of development within limitations and balance (i.e., sustainability) of the non-anthropocentric sphere. There are arguments of the romanization of the "primitive Man" (á la Illich (1983)) within Nature in which life was unquestionably hard with much shorter lifespans. With the recognition of arguments of over-romanization, the intensifying hubris of our control of all of Earth leads to <u>D</u>evelopment that is unbounded and unlimited. Within the world <u>D</u>evelopment is fueled by intensifying isolationism (e.g., through extreme nationalism, racism, homophobia, and xenophobia) and hubris of Earth's control, in which such ideologies must be taught for its continuing hegemony. As ecopedagogy is an enemy of <u>D</u>evelopment and hegemony, ecopedagogical spaces are to demystify and transform toward <u>d</u>evelopment as the word is uttered. On the contrary, hegemonic entities will always suppress ecopedagogy, as well as critical pedagogies overall, as countering their goals.

Democracy, in its widest of ideological framings, is also under attack when <u>D</u>evelopment is the endpoint. In one of my writings I wrote the following on this issue:

> Without understanding the injustices of many normative framings of development, acts of environmental ills are grounded in ignorance, severely hindering democratic practices by decisions, as well as "solutions," constructed from false, politically tailored knowledges. (Misiaszek, 2018b)

Teaching <u>D</u>evelopment structurally hides world-Earth truths that limit democratic decision-making to ideological truths that continue hegemony. Questioning possibilities of democratic participation in self-progress must also include questioning the knowledges on which one's foundations in decision-making are grounded. This is yet another area where the dialogical process of Freirean pedagogies is essential, to both understand others' knowledges, understandings, and experiences, as well as, specifically within this line of discussion, to deconstruct one's Self epistemological understandings. Ecopedagogues ground the problem-posing of ideologies as critical pedagogues, "the underlying epistemological and ideological assumptions that are made about what counts as 'official' or legitimate knowledge and who holds it" (Apple and Au, 2009; Apple, 2004, 2000).

This includes the emergences and dialogues around the conflict of diverse, often opposing, understandings which many will hold close to them. Freire

(2000, 1998d) and Apple (2004) have discussed extensively how such conflict is structurally avoided and falsely neutralized within schooling, a topic discussed earlier in this chapter and to which this book will return to throughout. I have written the next passage in another book in which I discuss that, with development, conflict will inevitably emerge from such discussions but Jickling and Wals (2008) remind us that

> "democracy depends on differences, dissonance, conflict, and antagonism, so that deliberation is radically indeterminate" (Jickling and Wals, 2008; Goodman and Saltman, 2002; Saul, 1995). . . . All environmental pedagogies will inevitably encounter these conflicts, but the difference in ecopedagogies is that conflicts are beyond not ignored, but that teaching is grounded at these points of tension, especially in the deconstruction of developments for the term's reconstruction that is within socio-environmental models. (Misiaszek, 2018b)

As described by the civil rights educator from the US southeast within Appalachia, Miles Horton stated the following in the "talking book" with Freire, *We Make the Road by Walking* (1990), on how democracy must guide development rather than the other way around.

> The more people participate in the process of their own education, the more the people participate in the process of defining what kind of production to produce, and for what and why, the more the people participate in the development of their selves. The more the people become themselves, the better the democracy. (Horton et al., 1990)

The wording of the "development of their selves" is not to be read as individualism within neoliberal framings as *outside* society but rather as *part of* society in the collective Self within collective "becoming themselves."

Ecopedagogy widens such decision-making within the sustainability of Earth; however, this does not contradict what Horton states but rather expands the focus of "defining" beyond the world. Teaching/reading the inseparability of world-Earth violence allows for better understanding in determining what is needed to be known to end social injustice, with "becoming[ing] themselves" very much connected to environmental violence. When false ideologies of world-Earth separability are peddled by banking-education models, the participation of one's own education—one's own right for democratic educational construction—is unjustly limited, at best, and falsified, at worst.

Outside of students-teacher(s) spaces, individuals' ecopedagogical literacy is purposely constructing self-understandings from a widened, diverse range of perspectives toward a better understanding development for "who" and the (dis)connections of sustainability for Earth holistically. Although there are various differences of in-person dialogue within traditional learning spaces, there are various ways to diversely bring others' arguments into such learning spaces. For ecopedagogical literacy, this includes reading one's Self, such as self-reflectivity of one's own ways of knowing to diversify one's own readings (i.e., through ecologies of knowledges). I would state this is sometimes the most important one, especially within our increasingly fast-paced, connected world online in terms of voices being able to "get out" without traditional, historical gatekeepers, such as in terms of people (i.e., gatekeepers themselves), structures, and costs. However, there is also the hidden structuring of post-truthism and neoliberal ideologies, among many other oppressive framings, within the distribution of false knowledges within social media, online search optimization, and so on. This is also within academia, within both direct and indirect censorship throughout the world, in which I have argued elsewhere that the former is easier to recognize than the latter, but both result in censorship (Misiaszek, 2018a). Teaching ecopedagogical literacy is for self-reflectivity and deconstructing the existence of these politically hidden structures, in which their strength is within their veilness. It is the vastness of people not questioning what is development and sustainable development that further entrenches dominant ideologies, even though they oppress the very persons who do not critically question them.

Ecopedagogical literacy focuses upon the socio-environmental violence connections within all spheres of and within Earth, as to first transform through questioning the unquestionable ideologies that are inherently oppressive, with the very relevant topic of critiquing "development." As stated at various times previously, and as will be restated many times again, schooling is a tool in which such critical questioning is either encouraged or discouraged. Gadotti and Torres (2009) wrote on Freire and development within the Brazilian context, arguing that colonial histories entrenched with banking education, if any education at all, has produced a "cultural backwardness" that did/does not allow for deep democracy or "authentic development" (termed *d̲*evelopment my publications). Freire (2000) explained that development has been framed by the oppressors rather than development grounded in ending oppressions.

> Deprived of their own power of decision, which is located in the oppressor, they follow the prescriptions of the latter. The oppressed only begin to develop when, surmounting the contradiction in which they are caught, they become "beings for themselves." (2000)

Gadotti and Torres (2009) argued that it is through ecopedagogies that development can emerge as development critiqued, along with previous schooling as a tool for entrenching development constructed by ones' oppressors, for contextualized authentic development to emerge. Ecopedagogical reconstruction of development toward authenticity is essential in balance (i.e., sustainability) with the non-anthropocentric sphere. This happens within the world as the inherent socio-environmental violence connections, expanding beyond Gadotti and Torres' (2009) initial arguments, but also coincides with Gadotti's (2008b, 2008a, 1996) other writings on Gaia and planetary citizenship education, and Torres' (2017, 2016, 2013b; Tarozzi and Torres, 2016) writing on critical GCE.

Before going on to the next chapter, there should be some discussion on coloniality and development, which will be and has been discussed throughout this book. Ecopedagogues must teach by problematizing coloniality that affect "the notions of development, progress, humanism, culture, representation, identity and education" (de Oliveira Andreotti et al., 2012) in sustaining/intensifying socio-environmental oppressions. As will be discussed later in this book, ecopedagogical spaces center deconstructing socio-environmental oppressions rooted in coloniality, within the coinciding and conflicting theoretical frameworks of post-, anti-, and neocoloniality across globalizations. Constructing true "postcolonial" planetary *d*evelopment, minimizing coloniality's influences to elimination, coincides with Ali Abdi's (2008) call for de-colonial teaching for praxis to utilize processes of globalization that counter colonial-structured education, and for ending globalizations that sustain and entrench neocolonialism.

The next subsection will further analyze these aspects in the conflicted term of "sustainable development." I will discuss the teaching to ecopedagogically read the foundations and constructs of "sustainable development" and its education (specifically written as "es*D*"/"ES*d*" to indicate what type of development is within the term, and "ESD" to indicate either).

2.4.1 Sustainability and education for sustainable *d*/*D*evelopment (ESD)

A key question with regard to "sustainability" or "balance" is as follows: What is environmental "violence," and how does this relate to social (in)justice? The latter part of this question is related to the fact that injustice does not emerge from the nonhuman non-anthropocentric sphere. Problematizing the source of self-reflected, purposeful violence from the world onto the non-anthropocentric sphere in the name of "development" is essential in ecopedagogies. Paraphrasing and somewhat reinventing the arguments of Dewey (1963) and Freire (2004, 2000), humans are the only ones that can reflect upon their violent actions through their histories, dreams, and widened analysis of beyond what is in front of their eyes currently. Figure 1.1 illustrates the deepened and widened connections and accompanying politics between social and environmental violence that we place onto the larger planetary sphere. There are various complications with this overlapping within many epistemological structures, with social violence of the world as part of violence against Earth holistically, and the world as part of Earth. Such constructions of environmental violence as the umbrella term for social violence counter anthropocentric groundings of determining actions. The term "violence," which we often view as only from, and done upon, humans, is expanded to include Earth, as part of the violence committed upon the anthropocentric as well as the planetary, non-anthropocentric spheres. Banking-education models suppress our "histories, our dreams, and [extend] widened analysis," as I wrote earlier, to only what is prescribed by the dictated ideologies of schooling, resulting in reproducing justifications for unsustainable socio-environmental violence with limiting "analysis" for ideological agreement and fatalism in possibilities and in reach/scope.

In addition to problematizing *d*/*D*evelopment, which the previous section focused on, we must problematize neoliberal "sustainable development" (S*D*) and its education (ES*D*) which are often falsely portrayed as neutral and apolitical. With the same letter casing as previously explained for *d*/*D*evelopment, this book utilizes ES*d* (and S*d*) as well as ESD (and SD) to indicate possibilities for either model. How is balance/sustainability defined and through which framings of "development" are they viewed, are important ecopedagogical questions for analyzing SD. To begin this section, we revisit some of the main factors of *D*evelopment that were discussed in the previous section:

> (1) economics [is] the sole factor of development analysis; (2) economic development without economic justice considerations by ignoring how development processes sustain/increase hegemony, widening the gaps between the economically rich and poor; and (3) local framings of development are disregarded for globally constructed ones "from above"—most often, but not limited to, Western development models. (Misiaszek, 2018b)

Ecopedagogical teaching and reading of sustainability is ingrained *within* and coincides *with* deconstruction and reconstructing—that is, development.

Encompassing "within" refers to the teaching and reading to better understand how development is defined within what is "sustainability," including what is the baseline for its goals, sustainability for whom, and the degree of inclusion of the non-anthropocentric sphere within the term's usage.

Ecopedagogical reading of ES*D* against ES*d* problem-poses what "should" be sustained. Sustainability of development for holistically transforming the world within social justice models and within the balance of the static laws of Nature frames "sustainability" within ecopedagogical tenets. Similar to critical pedagogy's goal of teaching for praxis to end oppressions, ecopedagogies' work is toward ending unsustainable violence of the non-anthropocentric sphere from the world. Returning to the question posed at the beginning the previous paragraph, are these tenets/goals the focus when answering "what should be sustained?"

In a previous work, I wrote that attempts at socio-environmental peace and planetary sustainability within neoliberal framings and pedagogies are "akin to pushing a round peg through a square hole" (Misiaszek, 2018b). Looking back, I think I was far too optimistic in this analogy. A better analogy would be "trying to drive a bulldozer through a shirt's buttonhole." It is possible and unfortunate that in a future publication I might need to revise this analogy again because I am being too optimistic.

Gadotti has written that sustainable development must "not only [be] attached to economic values, but also to cultural values that encourage solidary practices" (2008a). This quote further indicates the inherent connections of global solidarity to SD both in its scope—what should be sustained—and in that sustainability practices are contextual. Gadotti further argued that solidarity must also be planetary (utilizing constructs such as Gaia and planetary citizenship) and that SD must be global and planetary, but SD teaching and practices must be local (Gadotti, 2008a) (i.e., ES*d*).

Both of these can be linked to planetary citizenship and critical global citizenships, with the important notation on GC/E, once again, that citizenships can also be exclusion, oppression, and planetary unsustainability (Fanon, 1967, 1963; Memmi, 1991b; de Sousa Santos, 2018, 2012a), as well as models of *D*evelopment and sustainable *D*evelopment (Misiaszek and Torres, 2019, Misiaszek, 2020, 2018b; Gadotti and Torres, 2009; Gadotti, 2008a, 2008b). The positive and negative aspects of citizenships with *d*/*D*evelopment very much coincides with the contested terrain of the SD models of "who" is affected and how.

Sustainability *coincides with* development due to our "unfinishedness," which Freire (2000) discussed is unique to humans, and thus "development" corresponds only to being humans; however, *D*evelopment is innately dehumanizing and unsustainable. Human reflectivity for transformation is described by Freire (2000) as follows:

> [As] uncompleted beings, man is the only one to treat not only his actions but his very self as the object of his reflection; this capacity distinguishes him from the animals, which are unable to separate themselves from their activity and thus are unable to reflect upon it. In this apparently superficial distinction lie the boundaries which delimit the action of each in his life space. Because the animals' activity is an extension of themselves, the results of that activity are also inseparable from themselves: animals can neither set objectives nor infuse their transformation of Nature with any significance beyond itself. Moreover, the "decision" to perform this activity belongs not to them but to their species. Animals are, accordingly, fundamentally "beings in themselves." (2000)

How human unfinishedness connects to the rest of Nature is through their ability to reflect upon their actions and to devise development goals grounded on planetary sustainability, rather than the opposite. Although some scholars have identified this point as evidence of Freire devaluing all other living beings, Freire is rather discussing humans' obligation toward the world and later in his writings expanding this to include Earth (Gadotti and Torres, 2009; Misiaszek and Torres, 2019). In later works, before his death, which focuses on ecopedagogy (Misiaszek, 2018b; Gadotti and Torres, 2009), *d*evelopment only occurs when it coincides with socio-environmental sustainability, calling for re/deconstruct our subjective "development" to be balanced with Earth's static (or objective) laws of Nature. More in-depth discussion on the Freirean aspects of this will be taken up in the next chapter.

World-Earth relationship does have a balancing point; however, how our subjective world determines this point could be very different from Earth's true balancing point. Worded slightly differently, what is the baseline that defines sustainability or unsustainability? Ecopedagogical reading to determine this baseline is endlessly complex. This is especially true in an increasingly globalized world with unbalanced distribution of *d/D*evelopment practices and outcomes. However, if *D*evelopment through neoliberalism is increasingly universal, we are accelerating away from this baseline and balancing point. The wording of "unequal distribution" is to reemphasize ecopedagogies' overall ecopedagogical question of who benefits and who suffers from environmental violence, with the non-anthropocentric sphere as "suffering" from unbalanced violence. If we sustain a baseline of today's world (2020), current injustices and environmental devastation will continue without question, with large gaps of equal distributions that cause much of the world to continue suffering with sub/conscious ignoring of the non-anthropocentric sphere. This results in widening of the gap between false world baselines and Earth's true balancing point.

Let's first take up the issue of global warming if sustainability's baseline should be set at the current level, continuing from the previous paragraph. As most reputable research have found, we are already at or, more likely, over the tipping point, which strongly indicates our need to decrease the use of fossil fuels. Although this fact should dictate how we define sustainability's baseline, Development counters such logic with neoliberal illogicalness. Accompanying this is distribution and baseline questioning affected by globalizations. Giving the example of (un)sustainable distribution of natural resources, sustainability's baseline set at the current level could be accompanied by more unequal distribution among the world's societies. However, achieving such ideal equality in distribution does not, most often, necessarily correspond to Earth's balance or limitations. This emphasizes the fact that SD achievement in the world, even with equity and social justice, does not necessarily mean the processes are toward planetary sustainability.

Returning to SD within socially-justice models and aspects of "distribution," it is necessary to discuss the sociohistorical complications of "truly" achieving it. Determining if distribution should be equal or sociohistorically equable is central to ecopedagogical work. Equable disruptions of globalizations from above are essential for radically altering distribution and to account for reparations for sociohistorical oppressions, such as past colonialization that continue as neocolonialism. Such exponentially radical equity counters current hegemony built upon historical oppressions and thus would require tremendous transfers of power from the Global North to the Global South. This includes loss of epistemological hegemony from the epistemologies of the North, for example, factors historically grounded in colonialization, capitalism, and patriarchy, as de Sousa Santos (2018) suggests. Freire's (2000) arguments that class suicide is near impossible, begs the following question of who would be willing to give up their power. This question is especially important to ask when globalizations from above, neoliberalism, and ideologies of Western exceptionalism instill the thinking that current dominant world systems are "natural" and "unquestionable." These are only a few key complexities that emerge when determining sustainability's baseline.

It is important to note again that the last paragraph mostly included the difficulties of sustainability when the baseline is recognized at the current level of environmental violence. The increased complications of the aspects discussed are not linear but rather hyperbolic, in the necessary loss of hegemony required for true Earth sustainability. Critical pedagogies, including ecopedagogies, teach toward eliminating hegemony as unjust power but its accomplishment might be viewed as impossible—hegemony and neoliberal needs such fatalism to sustain. On the contrary, ecopedagogies are utopic with such global hegemonic loss necessary for world-Earth sustainability to be achieved as a result of learners' praxis. Such teaching is with the recognition of endless difficulties and complexities toward this goal. The next section will begin discussions on how individuals' and societies' livelihoods coincide and conflict with *d*/*D*evelopment and world/Earth sustainability.

2.4.2 Livelihoods read with *d*/*D*evelopment

Ecopedagogical reading of livelihood includes problem-posing the following two questions: (1) How does livelihood relate to our needs and "wants"? and (2) Who is included in our defining of livelihood, with the "who" including local-to-planetary spheres? Dialogue and reading around answering these questions also include the following question: What is the balance of livelihoods between the "Self," within one's self-defined societies, and "others'" societies? Here again we return to previous questions of what is "development" and whose "development," what are the baselines of "sustainability," and what is/are the connectivity/solidarity within and between spheres, possibly in terms of "citizenship." Juxtaposition with ecopedagogical questioning of development, an important question is the following: How does the gain of one's livelihood beneficially and/or negatively affect others' livelihoods? Ecopedagogical work would inherently counter livelihood framed as zero-sum goals, which would counter globally all-inclusive socio-environmental justice and planetary sustainability.

Postma's (2006) argument on neoliberalism's over-prioritization of the Self's private sphere has been discussed throughout this book and is especially significant to the topic of livelihood. Neoliberal's sole focus on decision-making with regard to economic gain and the associated power defines and measures livelihood of the Self and their private sphere (Postma, 2006). This is partially true for almost any construct of livelihood, but it is the singleness of its priority within neoliberalism that inherently counters concerns of social justice outside of one's private sphere, while ignoring all other wider spheres unless they affect one's private one, which counters socio-environmental justice and sustainability (Misiaszek, 2018b, 2011).

As discussed previously, ecopedagogical readings of economics are important as contested terrains, but ecopedagogical work inherently counters neoliberal ideologies without question. Ideologies that have actions for environmental well-being as reversely related to livelihood must be problematized, with the inclusion of critical pedagogies countering the same inverse relationships between livelihood and social justice, especially social justice for the entire world sphere. If livelihood is within neoliberal terms of ever-accelerating consumption without end, neoliberal livelihood inherently has goals that are unsustainable and entrenched with social injustices/violence. Ecopedagogical literacies are unceasing in counter-reading livelihood models which work to these ends. This sentiment is eloquently written by Freire (2000) as follows:

> The oppressors develop the conviction that it is possible for them to transform everything into objects of their purchasing power; hence their strictly materialistic concept of existence. Money is the measure of all things, and profit the primary goal. For the oppressors, what is worthwhile is to have more—always more—even at the cost of the oppressed having less or having nothing. For them, to be is to have and to be the class of the "haves."

Within the United States, and in many other parts of the world, the expression "pursuit for happiness" is often invoked as (in)directly connected to livelihood. However, this can be contextually problematic. I have experienced that the mention of "livelihood," and much more so "lifestyle," invokes anger in the Global South as billions are just barely surviving, often associating livelihood as a Western concept created by those who largely make up the "haves" compared to the "have nots." Within such dire circumstances there is the underlining apathy within neoliberal ideologies that calls for individualized happiness for a minority without regard to the pursuit of sheer survival as a result of environmental violence-connected oppressions. Within neoliberal-constructed livelihood, it is the private sphere that alone is valued (Postma, 2006), so survival of the world's majority is inconsequential. In other words, the survival of "others" is ideologically perceived as unimportant as one's own private sphere livelihood gains are paramount. Without teaching to radically change such normative neoliberal ideologies in caring for the world and the widened Earth holistically, I argue, once again, that environmental pedagogies will fail.

I end this section on the note that global warming further heightens the need to problematize livelihood for a few inversely affecting the survival of many others. With 2 to 25 million people expected to become environmental migrants/refugees over the next few years (International Organization for Migration

(IOM), 2014), this population's survival, including these societies' land, is taken away from them for others' livelihood. The question of whose de-*d*evelopment is done for others' *D*evelopment also needs to be asked within larger societal spheres. There will be additional discussions on environmental migration in the last chapter.

Notes

1 The writing of "the fish" and "it(self)" here and elsewhere is not perfect, as it objectifies fish, but allows for better readability. Such objectification of nature is present in the writing in other parts of this book. This issue highlights the limitations and epistemological challenges of the English language that often inherently objectifying Nature, calling for the need of ecolinguistics (briefly discussed later in this book).

2 The term "acquiring" here is not necessarily through technical aspects of education, such as through banking-education pedagogies, but rather refers to all pedagogies of constructing "knowing," including critical, Freirean pedagogies with formal, nonformal, and informal models.

3 The terms "who/what" is given to signify a biocentric framing of contextualizing globalization that includes not only humans but also all other life beings and the nonorganic natural world (e.g., landscapes, seascapes).

4 "The intensification of worldwide social relations which link distant localities in a way that local happenings are shaped by events occurring many miles away and vice versa" Giddens (1990).

5 In the context of including Freire's quote, it is important, and telling, to include the following footnote he had for this statement in his book.

> To this end, the invaders are making increasing use of the social sciences and technology, and to some extent the physical sciences as well, to improve and refine their action. It is indispensable for the invaders to know the past and present of those invaded in order to discern the alternatives of the latter s future and thereby attempt to guide the evolution of that future along lines that will favor their own interests. (Freire, 2000)

6 At three feet, though, a tide of blue covered Hudson River Park and West Street. Four feet, five feet:

> The blue crept east along Canal, toward the entrance to the Holland Tunnel. At six feet, . . . the blue covered parts of La Guardia and JFK airports, the Williamsburg waterfront, Roosevelt Island, and Brooklyn Bridge Park. . . .

even the supposedly manageable increase of 1.5 degrees Celsius envisioned by the Paris Agreement would translate to around ten feet of eventual sea-level rise. When I clicked up to ten feet, much of Battery Park City, the Lower East Side, and Brooklyn's waterfront was submerged. The Dumbo carousel stood solitary in the East River, and the barrier spits of the Rockaways and Coney Island mostly vanished. The program also has a function that allows you to see the outcomes of greater increases in temperature. At 2 degrees—equivalent, in Climate Central's estimation, to about 15 feet in sea-level rise—the water completely surrounded the pools of the 9/11 memorial. At 3 degrees—20 feet—the water overwhelmed them. I clicked up to the maximum setting of 4 degrees—30 feet—and maneuvered upward to take in the view from the top of the spire of One World Trade Center. Lower Manhattan had become an archipelago, and the rooftops of southern Brooklyn resembled boats bobbing in a marina (Rice, 2016). This is New York in the not-so-distant future.

7 This statement tries to acknowledge "citizenship" beyond the term itself because "citizenship" is often criticized for being a Western construct and, too often, a tool of othering "us" (i.e., fellow citizens) from "them" (i.e., noncitizens). This will be discussed later.

The term "productivity" here is not utilized within neoliberal and/or capitalist ideologies, but rather how individuals become productive members with societies in a wide range of social roles. However, there are neoliberal citizenship models that will be discussed later, as well as notions of the citizen as the consumer.

8 Although I problematize Illich's writing as gendered (i.e., "primitive/modern man"), there is the argument that maybe, to some degree, it is appropriately gendered due to socio-environmental oppressions due to patriarchy. Orally, I have argued that the gendered primitive and modern "Man" might not be negative as gendered because it is often men historically who have furthered oppressions, injustices, and unsustainability compared to their female counterparts. However, it is probably best if this was contextual, and the all-inclusive gendered masculine wording should be widened to "primitive/modern humans."

Part Two

Foundations of Ecopedagogy

3

Freirean Reinventions

Ecopedagogy

Already in this book there have been numerous references to Paulo Freire's work; however, this chapter will solely discuss many of the key aspects of Freire's scholarship for ecopedagogical reinventions of his work. It must be noted that it will be incomplete in full thoroughness due to this book's limited space and that there are endless reinventions, including ecopedagogical reinventions, of his work. In other words, Freire's work is extremely complex, voluminous, and contextually reworked in limitless ways in which I will discuss many aspects, but it will be far from being all inclusive. This chapter begins with a brief history of Freire on environmentalism, a topic which he has been criticized for not addressing, although it was to be the topic of his next writings which, unfortunately, he was not able to complete due to his death in 1997 (Misiaszek and Torres, 2019). From here the chapter delves into four Freirean elements as foundational to ecopedagogies: creating learning spaces that help determine the politically hidden oppressor/oppressed socio-environmental relationships; democratic/horizontal learning spaces in which students and teachers both teach and learn from one another; pedagogies of hope, rather than fatalism that emerge from neoliberal education models; and praxis-based teaching toward transformation. Some of these have been already discussed at some length, but the sections in this chapter will further detail and exemplify these aspects.

Before continuing I want to emphasize the Freire notion of reinvention briefly here and more later, to help counter possible critiques that this book, my other ecopedagogical work, and others' ecopedagogical work on this topic is manipulating Freire's work. A key construct of Freire is reinvention, in which he did want his work to be repeated throughout the world but rather be reinvented contextually within the society and with time. His concepts of reinvention have been analyzed in various pieces of his writing, but I would argue this through highlighting his arguments on humans' "unfinishedness"—a topic to be discussed later in its own section—that includes their constructs of pedagogies. Unfinishedness in that pedagogies are grounded in their essence but are never complete but contextually adapting, with the recognition of diversity and the unfinishedness of individuals and societies. Ecopedagogy emerges from reinvention of Freirean pedagogies and scholarship including and beyond the field of education, reinventing the essence of his work within education for ending environmental violence that leads to social oppressions. He wrote some on this topic throughout his work, especially his later writings (Misiaszek and Torres, 2019; Gadotti and Torres, 2009), so reinvention also includes utilizing his work on the essence of ecopedagogies, as he never used the word "ecopedagogy" directly. Ecopedagogues also reinvent his work to environmentalism beyond the human realm (or beyond the anthropocentric realm into focus upon the non-anthropocentric sphere also).

3.1 Freire as an environmentalist: A very brief introduction

> Freire was talking about ecopedagogy. In an interview at the Paulo Freire Institute, he talked of his love for the Earth, the animals, the plants: "I want to be remembered as somebody who loved the men, the women, the plants, the animals, the Earth," he said on that occasion. In one of his last books, A Sombra desta Mangueira (literally, In the Shadow of this Mango Tree, published in English as *Pedagogy of the Heart*, 1998), he speaks of the pleasure of breathing pure air, the joy of entering a river that has no pollution, of stepping on grass, or the sand on the beach. He criticized the capitalist logic that gives no value to those free pleasures, and substitutes for them the pleasure of profit. Capitalism substitutes the free happiness of satisfying human needs for happiness that can be bought or sold, but above all, for satisfying the needs of capitalism—not human needs, but needs imposed upon human beings by the search for profits. Freire did not separate human needs from the needs of the planet. When he died, Freire was writing a book about ecology. (Gadotti and Torres, 2009)

Freirean reinventions of critical environmental pedagogical models (e.g., ecopedagogies) centralize the problem-posing socio-environmental issues for authentic understandings for genuine freedom, participatory citizenship, and democracy (Gadotti 2008a,b,c). In personal conversations with Moacir Gadotti, Carlos Alberto Torres, and a few others, they stated that Freire's next book was to be on ecopedagogy and also how Freire discussed how a chapter on ecopedagogy was a needed missing chapter from *Pedagogy of the Oppressed* (Misiaszek and

Torres, 2019). Unfortunately, Freire passed away before being able to complete this book on ecopedagogy or add it to an updated edition of his most famous book. Carlos and I wrote Chapter 5 (Misiaszek and Torres, 2019) in Carlos' *Wiley Handbook of Paulo Freire* (Torres, 2019), as if Freire had written it himself. In the chapter we expressed humbly our inability to replicate Freire and admitted that our words would never be the same or as good as Freire's.

Environmentalism is one of many topics in which Freire has been critiqued for not extending beyond issues of class, into, for example, topics such as gender, race, and sexual orientation, among others (Au and Apple, 2007). With environmental issues, Freire did write some powerful words, especially in his later writings, as given below and other passages which are scattered throughout this book.

> I do not believe in loving among women and men, among human beings, if we do no become capable of loving the world. Ecology has gained tremendous importance at the end of this century. I must be present in any educational practice of a radical, critical, and liberating Nature. (Freire, 2004)
>
> The notion seems deplorable to me of engaging in progressive, revolutionary discourse while embracing a practice that negates life—that pollutes the air, the waters, the fields, and devastates forests, destroys the trees and threatens the animals. (Freire, 2004)

Freire discussed the inseparability between social and environmental justices, as pedagogies are constructed for transformation through "educational practice[s] of a radical, critical, and liberating Nature" (Freire, 2004). The second quote elaborates on specifics outside of the anthropocentric sphere to include the organic sphere, but also the inorganic with "the waters, the fields." It should be noted again that in this book and past publications of mine (Misiaszek and Torres, 2019), I (along with Torres) have changed his wording of "the world" in the above quote to become "Earth"—as I have already done up to this point in this book. I/we have utilized this terminology from Freire's (2000) own words in describing a discussion with a peasant on his response to there would be no world without human beings and the social roles between them (Freire's quote given later in this chapter).

Within this terminology, Freire is calling for the need to love Earth as we love the world—as teaching that embraces love between the world-Earth sphere. This book delves into, as our missing fifth chapter (Misiaszek and Torres, 2019) attempted to do to a limited extent, the epistemological complexities that emerge from love of the world holistically and Earth in which the world is a part of.

As Freire's work overall represents masterful, detailed deconstructions of the arguments on education and sociological aspects overall, Carlos and I (Misiaszek and Torres, 2019) also argue that his work on ecopedagogy, if not for his untimely death, would have had the same completeness in deconstruction and reinvention.

3.2 "Unfinished" world and "Finished" rest of Earth

There are various ways in which I could begin the topic of the "unfinishedness" of human beings, a topic which grounds much of Freire's work, especially with arguments for education to be utopic rather than fatalistic. Freire wrote how this is one of the most devastating aspects of neoliberalism. I will begin this discussion using Freire's words on the differences between transformation, adaption, and development below.

> While all development is transformation, not all transformation is development. The transformation occurring in a seed which under favorable conditions germinates and sprouts, is not development. In the same way, the transformation of an animal is not development. The transformations of seeds and animals are determined by the species to which they belong; and they occur in a time which does not belong to them, for time belongs to humankind. . . . Women and men, among the uncompleted beings, are the only ones which develop. As historical, autobiographical, "beings for themselves," their transformation (development) occurs in their own existential time, never outside it. (Freire, 2000)

An important aspect of ecopedagogical work is differentiating between "development" and "transformation," as I have explained to some extent already. This comparative argument should be focused on the framings of the terms, rather than the usage of the specific terms as they are plural in meanings/framings. In another words, scholars will define "development" and "transformation" differently in that the arguments I provide here might be reversed in terms, but not in the essence of the framings. Freire discussed the differences between reflective development and transformation as defined by change without reflection—opposite to praxis. Banking education leads to transformation rather than development, and noncritical environmental pedagogies lead to transformation without environmental and sustainability concern rather than development within socio-environmental justice and planetary sustainability.

Freire also differentiated between "living" and "existing" to "not only live but exist" (Freire, 2000),[1] which I would argue are "development" and "transformation," respectively. All these arguments state the differences between humanizing and dehumanizing pedagogies that respectively coincide with development and existing, rather than the later transformation (without development) and living. This is elementary to ecopedagogical work focusing on the world to end socio-environmental injustices, but also for widening this focus to the planetary sphere for sustainability and non-domination of the non-anthropocentric sphere (i.e., all of Earth).

Once again, I return to the complexities and often the conundrums of the world's subjectivity that coincide and conflict with Earth's laws of Nature, within the fact that the world is part of the larger Earth. Adaption within the laws of Nature is how the rest of Earth returns to balance/sustainability, with humans as the only beings that can purposely challenge it (i.e., having the ability to understand that our actions are ones of unbalance/unsustainability). But does increased understanding of this have made the world more sustainable? Invoking Illich's (1983) work again between the primitive and modern man, both "men" know they are challenging Nature, but with the former knowing the risks and the latter believing "he" can challenge Nature without consequences. Individual and societal unfinishedness allows for there to be constant development; however, as discussed previously, the question is how to define and balance development that causes environmental violence with the rest of Earth and end social injustices and de-*d*evelopment, from others' *D*evelopment. Freirean teaching through the aspects of unfinishedness provide utopic ideologies of possibilities, but it is not untethered. Utopic *d*evelopment is inclusive of not denying others' unfinishedness possibilities, others' *d*evelopment, or Earth's sustainability.

Freire expressed the need for teaching through the belief in one's unfinishedness, rather than banking-education models that teach fatalistic ideologies, as individual and societal finishedness—one of the many reasons Freire opposed banking education (2004, 2000, 1998b, 1992). This is not to say that all Freirean critical teaching with ideologies of unfinishedness will lead to (re)invention that ends social injustices and/or unsustainability, but structurally encourages possibilities for transformation rather than systematically denying such possibilities. Banking education—which is interconnected with reproductive education—reproduces ideologies, including sustaining and often intensifying oppressions, without questioning normative, dominant ideologies. Finishness is within the constraints of these dominant ideologies, suppressing all development outside of them. Ecopedagogues teach to identify and counter oppressive ideologies and their fatalistic finishness by problem-posing the ideologies that lead to socio-environmental oppressions and unsustainability.

Within my wording in this book, this includes teaching for praxis toward either ecopedagogical development or reproductive Development.

Banking environmental education teaches for Development, as critical reflection, creativity, and curiosity is structurally limited, if not eliminated, for transformative praxis to not occur. Returning to Gadotti and Torres' (2009) article on how the cultural backwardness of Brazil, as "originat[ing] from its very history [and] thus, the possibility of elaborating a new Brazilian pedagogy conforms perfectly with the analysis of this concrete process and the detection of the cause of those cultural complexities that block the way to freedom." Such backwardness has emerged from banking-education models not leading to transformation or thick democratic societies. Praxis will be discussed later at length in this chapter, but within these arguments here it is important to express briefly on how Freire (2000) wrote on praxis being societies' criticalness emerges deepened and widened reflection of actions.

> People will be truly critical if they live the plenitude of the praxis, that is, if their action encompasses a critical reflection which increasingly organizes their thinking and thus leads them to move from a purely naive knowledge of reality to a higher level, one which enables them to perceive the causes of reality.

The "plenitude of the praxis" cannot occur within fatalistic education models that teach ideologies that transformation is impossible, reproducing the same future as the past and present. In other words, without possibilities of "better futures" through utopic reflection, reflections of actions are limited to current systematic socio-environmental oppressions and planetary unsustainability practices.

Ecopedagogues focus on unfinishedness of the world, because of how our actions affect the overall Earth, especially those toward (in the utopic sense) or away from sustainability. Finishness is the essence of the non-anthropocentric sphere, in that "finishness" refers to the static balancing point of Nature and the true objectivity of the laws of Nature. Unfinishedness is unique to humans, which highlights the differences of development of the world and adaption of the non-anthropocentric sphere, but Earth (excluding the world) is unfinished as it evolves and changes, such as rocks turning into sand over time.

As previously mentioned, Freire did write on the aspect that our histories do construct our futures; however, as indicated by the plural form of "futures," Freire did not mean this in a fatalistic way. Below Freire (2004) uses the metaphor of building a wall in which "tomorrow" does not have to be a repletion of "today," or yesterday, but that it is a brick in the wall nonetheless.

> I must now return to a point always present in my reflections—my refusal to understand history as a determination) thus, my rejection of an inexorable tomorrow. Tomorrow is neither a necessary repetition of today, as the dominant would like it to be, nor something predetermined. Tomorrow is a possibility we need to work out and, above all, one we must fight to build. What takes place today does not inevitably produce a tomorrow.

Freire continues with the aspect of globalization in this quote, which will be discussed in another subsection. This quote could also be thought of within ecopedagogical spaces of what is development, in asking what *d/D*evelopment are we progressing toward—brick-by-brick, metaphorically—to "produce a tomorrow." Opposite to banking model environmental pedagogies, sociohistorical analysis is essential within ecopedagogical spaces with the understandings that reproduction of the past is essential—within the same ideologies. In short, it is sociohistorical analysis for transformational praxis rather than reproduction.

As opposed to the rest of Earth, it is only us as humans that make up the world in which we change through reflections of our histories, with histories constructing possible futures (plurality of futures indicating the opposing of fatalism). The two quotes by Freire below expresses these sentiments.

> Critical thinking contrasts with naive thinking, which sees "historical time as a weight, a stratification of the acquisitions and experiences of the past," from which the present should emerge "normalized" and "well-behaved." (Freire, 2000)

> Problem-posing education affirms men and women as beings in the process of becoming—as unfinished, uncompleted beings in and with a likewise unfinished reality. Indeed, in contrast to other animals who are unfinished, but not historical, people know themselves to be unfinished; they are aware of their incompletion. In this incompletion and this awareness lie the very roots of education as an exclusively human manifestation. The unfinished character of human beings and the transformational character of reality necessitate that education be an ongoing activity. (Freire, 2000)

It is problematizing, including the normative ideologies within our epistemologies, that provides the foundations for determining our "processes of becoming." In short, unfinishedness is humanness, with the aspect that this unique capability allows for us to better understand the rest of Earth, including how we are part of Earth rather than outside of Earth, as an objectification of Earth as separate for "our" explanation toward *D*evelopment without concern and reflection.

For praxis toward planetary sustainability, it is essential to problem-pose within ecopedagogical spaces the limitations of "our" unfinishedness within the rest of Earth's finishness. This is a question of rethinking/reimagining what is the non-anthropocentric sphere by deconstructing our sociohistorical relationships within the world as guiding our utilization of Earth's resources to fulfill *d*/*D*evelopment, all within the constructs of (un)finishness. Unfinishedness remains a foundation of ecopedagogies as Freirean pedagogies; however, the focus of non-fatalistic *d*evelopment within world-Earth sustainability (i.e., (un)finishness relationships) also grounds ecopedagogies. This can be seen as somewhat of a redundant statement because *d*evelopment and world unfinishedness with socio-environmental justice models already has these foundations within their constructs/framings.

3.2.1 Hope: Countering fatalism

> The ability to observe, to compare, and to evaluate, in order to choose, through deciding, how one is to intervene in the life of the city and thus exercise one's citizenship, arises then as a fundamental competency. If my presence in history is not neutral, I must accept its political Nature as critically as possible. If, in reality, I am not in the world simply to adapt to it, but rather to transform it, and if it is not possible to change the world without a certain dream or vision for it, I must make use of every possibility there is not only to speak about my utopia, but also to engage in practices consistent with it. (Freire, 2004)

Hope is the underlying aspect for the utopic and unfinishedness groundings of ecopedagogies from which praxis emerges, though transformative praxis can happen without hope. Banking environmental pedagogies teaching ideologies of only "adapt[tion]" within the current world of unsustainable environmental that is essential for *D*evelopment and *D*evelopment have no alternatives (or the finishness of development).

Hope is essential for radical praxis for transformative change within normative societal structures, which are often fatalistically taught as unquestionable and unchangeable. Students should be taught to freely dream of environmental utopias that counter current social, economic, and political structures. Without hope, transformation is limited within current societal systems, including oppressive systems, that purposely suppress the understanding of socio-environmental oppression and critical analysis abilities. Within ecopedagogical spaces there is a focus on hope of societal change of unsustainable environmental violence that affects the world and the non-anthropocentric sphere. The next subsection on globalization will analyze the non-/teaching of hope and accompanying possibilities of radical environmental as it affects more local spheres as processes from above and below.

Hopelessness to end Development ideologies is essential for hegemony's survival, but so is teaching that denies the questioning of actions for Development through ignorance, for continuing unrestrained environmental violence with minimal resistance. Without hope for development outside of Development framings, education for development will continue reproducing socio-environmental oppressions and planetary unsustainability. It is also environmental teaching without hope that continues teaching Development as unquestionably "natural."

Freire very clearly stated that "faith in humankind" (2000, p. 90, 91) within teaching is essential for it is "through conscientization that subjects assume their historical commitment in the process of making and remaking the world, within concrete possibilities, also making and remaking themselves" (Freitas, 2012). "Faith" of the world in ecopedagogical work is teaching through hope for environment transformation within the ecopedagogical tenets described previously. Freire's (2004) passage here describes, in this book's terminology, hope for world filled *with* development rather than Development.

> If it is possible to reach water by digging up the ground, if it is possible to decorate a house, if it is possible to believe this or that truth, if it is possible to find shelter from cold and heat, if it is possible to alter the course of rivers and to build dams, if it is possible to change the world we have not created, that of Nature, why not change the world of our own creation, that of culture, of history, of politics?

Utopia, hope, and faith are grounded in current realities, but also possibilities of new realities. Utopia and education, as Freire framed them, are not fanciful without understanding the realities, but are rather a horizon.[2] It is determining the utopic "horizon" of achieving socio-environment justice and sustainability that informs praxis, with critical reflectivity of the many barriers of processing toward that horizon (i.e., limit situations). Later in this chapter I will discuss utopic education through generative themes and problematize the barriers of limit situations, as termed by Freire; as well as problematize the United Nations' SDGs as utopic rather than fully achievable "horizons."

Ecopedagogical teaching and reading development and sustainability is problematizing taught future(s) to determine the limit situations that do not allow for a utopic world-Earth. Environmental pedagogies without such problematizing, or those that suppress it, are dehumanizing and deplanetarizing. Freire discuss this aspect of needing to problematize the future(s), which I would argue is problematizing "development" as the goals and processed toward those goals for futures.

> Deproblematizing the future, no matter in the name of what, is a breaking away from human Nature, which is socially and historically constituted. The future does not make us. We make ourselves in the struggle to make it. (Freire, 2004)

Without such problematizing, development and the future(s) that it provides coincides with dominant ideologies. I would argue with Freire that without such problematizing we are trained as nonhuman animals, rather than "intervening in the world," as he described below.

> Awareness of the world, which makes awareness of myself viable, makes unviable the immutability of the world. Awareness of the world and awareness of myself make me not only a being in the world, but one with the world and *with* others. It makes me a being capable of intervening in the world and not only of adapting to it. It is in this sense that women and men can interfere in the world while other animals can only touch it. That is why not only do we have a history, but we make our history, which equally makes us, and thus makes us historic. (Freire, 2004)

I would emphasize the need for recognizing our "intervening" as humans upon the rest of Nature with solidarity as being part of Nature/Earth.[3] Without being able to read development and livelihood through comparative problematizing of futures with <u>d</u>evelopment practices to othering/dominating <u>D</u>evelopment practices, we (i.e., humans) become disconnected from the world-Earth. Such problematizing includes how future(s) will coincide or conflict with "sustainability" of the subjective world <u>with</u> sustainability dictated by the static laws of Nature. Without wanting to be too grim, there is the fact that when the latter sustainability is ignored, as within neoliberal globalizations and neocolonialization, there is only one fatalistic future of the world's destruction through Earth's temporary unsustainability.[4]

3.2.2 Globalizations and fatalism

Although Freire (1998b) would not be alive after 1997 to witness the increasingly negative effects of globalizations from above (i.e., from "globalizers" to "the globalized"), as well as globalization from below, he wrote the following on, overall, globalization theory.

> Globalization theory, which speaks of ethics, hides the fact that its ethics are those of the marketplace and not the universal ethics the human person. . . . It

> is for these matters that we ought to struggle courageously if we have, in truth, made a choice for a humanized world.

Clearly, Freire here is speaking of globalizations from above, rather than possibilities for processes from below toward a more "humanized" and "planetarized" world, with the specifics of neoliberal globalization. There are various arguments Freire wrote on globalizations which mostly tended toward global markets as foundationally oppressive structures that systematically hide lack of care for humanity outside of economic profit that can be squeezed out of the oppressed. It could be argued that there is most often less caring for the rest of Earth (i.e., the non-anthropocentric sphere).

This section will focus on Freire's arguments of fatalism intensified by globalization with hope constrained by global neoliberal ideologies in which true hope cannot exist. Freire describes that education has become akin to training (i.e., technocratic education) for us to become cogs within the global market.

> If there are no social classes any longer, and if their conflicts are gone as well; if there is no ideology any longer, from left or right; if economic globalization has not only made the world smaller but made it almost equal, the education needed today has nothing to do with dreams, utopias, or conscientização (the building of critical awareness and conscience). In that view, education has nothing to do with ideologies) but rather with technical knowledge. It will be all the more effective if it trains the learners in certain skills. Introducing the dream of liberation, the utopia of social justice, into the teaching and learning of mathematics or physics or in the skills in training of factory workers, is to repeat ghastly mistakes for which we are paying a high price. (Freire, 2004)

Deconstructing global influences upon societal systems within education is essential to ecopedagogical analysis for teaching, reading, and research, as discussed in the previous chapters.

Because globalizations form contested terrains that can be either oppressive or empowering, with the key importance of how the processes affect local-to-planetary spheres. I reintroduce my argument here that ecopedagogical work on globalizations is not terminating all globalizations but ending processes of oppressions and unsustainability, and using globalizations that counter them. Freire wrote on this in the below quote discussing that taught false fatalism through globalizations from above are unchangeable and denies utopic possibilities, as I discussed in the previous section.

> Globalization is inevitable. Nothing can be done about it. It must happen because, mysteriously, that is how destiny has arranged things. So, we must accept what

> in essence only strengthens the control by powerful elites and fragments and pulverizes the power of the marginalized, making them even more impotent. Prisoners of fate. There is nothing left to do except bow our heads humbly and thank God that we are still alive. Thank God. And perhaps globalization too. I have always rejected fatalism. I prefer rebelliousness because it affirms my status as a person who has never given in to the manipulations and strategies designed to reduce the human person to nothing. The recently proclaimed death of history, which symbolizes the death of utopia, of our right to dream, reinforces without doubt the claims that imprison our freedom. This makes the struggle for the restoration of utopia all the more necessary. (Freire, 1998b)

Freire's discussions on the uniqueness of humans as historical beings, as well as, I would argue, his being influenced by Franz Fanon's (1967, 1963) theories of decoloniality, expresses, in this quote, the connections of fatalism and loss of one's history as dehumanizing education that counters one's innate freedom. Being still "alive" physically is not the same as "living" (wording from Freire's quotes previously given), with globalizations that further "fragments and pulverizes the power of the marginalized, making them even more impotent." Teaching for praxis counters such dehumanizing oppression to embrace globalizations that empower the marginalized and lead to planetary sustainability, which also includes how to end globalizations that oppress the marginalized and dominate/destroy the non-anthropocentric sphere.

With the sociohistorical aspects discussed here, I return to Freire's quote given previously in this section in which he focuses on the "produc[ing] a tomorrow" (2004) as transformative rather than reproductive. The below quote places this "tomorrow" within globalizations, with particular attention to technologies and the economy.

> The globalization of the economy or technological advances, for example, are not in themselves defining of a tomorrow given as certain, a sort of improved extension of a certain expression of today. Globalization does not put an end to politics, rather it creates the need to engage in the latter differently. While globalization may tend to weaken the effectiveness of strikes in the struggle of workers, it does not mean the end of the fight. The end may come to a particular form of fighting, striking, but not to the fight itself. It is tl1.en up to workers to reinvent how they fight, and not to settle before a new power. In fact, the inefficacy of strikes, at least in some sectors of the economy, must be understood quite broadly by progressives. What is at play in this inefficacy—which cannot represent, as the powerful suggest, the end of labor's struggle—is not simply the presence of technological advances that make globalization

> viable, but also the political use of those advances on the part of economic interests. That inefficacy represents a moment in the struggle and not its demise. In other words, nothing that takes place is the result of fatalism, but rather historical possibility. (Freire, 2004)

Technologies and economies have (non)problematically "advanced" dramatically since he wrote this passage; however, his overall question is more relevant today than when he wrote it: Have globalizations moved us more toward a single ideology or multiple ideologies that reproduce sociohistorical oppressions (i.e., toward fatalism) or possibilities for bottom-up empowerment? Freire's arguments on globalizations haven't reduced—and far from eliminated—the politics of education, globalizations have complicated the politics of the local and its education. Reinventing his work, ecopedagogues also center planetary sustainability within such problematizing of globalizations.

As I previously argued in this book and elsewhere (Misiaszek, 2018b), teaching of economics within ecopedagogical spaces is essential through critical lenses that radically counter dominant global neoliberal ideologies. The following quotes indicate the Freirean essence of this in reinvented ecopedagogies, with the second quote being planetary in scope.

> I reject the notion that nothing can be done about the consequences of economic globalization and refuse to bow my head gently because nothing can be done against the unavoidable. Accepting the inexorability of what takes place is an excellent contribution to the dominant forces in their unequal fight against the "condemned of the earth." (Freire, 1997)

> I imagine that easy living ranks high in their value system, one where a higher ethic, the one that rules the day-to-day relations among people, will have been almost completely nonexistent, replaced instead by the ethics of markets, of profit. According to it, people are worth what they make in money every month, and embracing the other, respect for the weaker, a reverence toward life-human, animal, and vegetable—a caring attitude toward things, a taste for beautifulness, the valuing of feelings—all this is reduced to almost no importance or to no importance at all. Although none of that, in my judgment, makes those agents of cruelty any less responsible, the fact in itself that this tragic transgression of ethics has taken place warns us how urgent it is that we fight for more fundamental ethical principles, such as respect for the life of human beings, the life of other animals, of birds, and for the life of rivers and forests. (Freire, 2004)

Freire very powerfully expressed the unsustainable socio-environmental violence that inherently emerges from fatalistic global neoliberalism, especially when unchallenged. Teaching that accepts injustices and unsustainability is fatalism characteristic of banking-education models. Technologies as "advancing" the world must be problematized in determining their usages from the local to the global sphere, such as problem-posing how some "benefit" while many others "suffer" from them. What must also be continuously questioned and re-questioned are the "endpoints" that dictate technologies' path(s). The singular or plural form of "path" indicates the problem-posing if technologies, and coinciding economies (and vice versa), are a single dominant global one or if globalizations, in some/many ways, diversify what we call "technology advances" and economies as we define "developed" or "developing" societies.

Returning to Freire's previous quote given on blindly adopting neoliberalism similar to jumping on a train without knowing its destination (Freire, 1998b), blindly following, creating, and using "advancing" technologies is aligned with this quote. Also, as Richard Kahn (2010) has argued ecopedagogues must determine the possibilities of radically shifting usages of technologies that are often used as tools of violence, oppressions, and unsustainability for ecopedagogical "re-tooling." Within Freire's (2004) own words, how can we transform economies and technologies, as processes of globalization, to counter "reduc[ing] the human person to nothing... proclaim[ing] death of history, which symbolizes the death of utopia, of our right to dream, reinforces without doubt the claims that imprison our freedom." The internet is one of the, if not the only, important tools of globalizations which allows for anyone to access information, including ways of knowing.[5] Does the internet diversify epistemologically how we "know," what we "know," or rather solidify epistemologies that reproduce oppressions and planetary unsustainability? As Kahn (2010) argues, equally as important is how we can transform the internet and the use of the internet for the former rather than the latter.

If globalizations are to make the world metaphorically smaller, how can tools of globalization be deconstructed and reconstructed toward ending oppressions and planetary sustainability within increased meaningful interconnectedness, while not devaluing diversity of individuals and societies. Boaventura de Sousa Santos (2018, 2014) would stress the need for ending paths leading to epistemicide, including our (i.e., academic scholars) role of analysis through sociologies of absences, this will be discussed at more length in Chapters 4 and 5. Freire's quote discusses education that teaches globalizations "such as it is" must be countered. I would argue with Gadotti (2008b, 2008c, 2008a) and Kahn (2010) that sustaining current globalizations from below and focusing on possibilities of reinventing other globalizations is essential for ecopedagogical work.

> In fact, the fatalistic discourse which says, "Reality is what it is. What can be done?" declares human impotence and suggests patience and astuteness for better adapting to life as an untouchable reality. Indeed, that discourse is one that sees history as determination. Globalization, such as it is, is inexorable. There is nothing to be done against it, except for waiting, until the very democracy that discourse has been ruining can remake itself, quite magically, in time to detain its destructive effect. (Freire, 2004)

Globalizations have the potential to have collectiveness of the world that holistically promote equality, equity, richness of diversity, and thick democracy, as well as providing the tools to work together for planetary sustainability. For example, as discussed previously, the internet allows for worldwide collective work to better understand and provide the pragmatic communications toward these ends. However, globalizations from above and Freire's point of globalizations not countering the current "as it is" reject such possibilities. Ecopedagogues very much align themselves with teaching to better understand and deepen democracy for world-Earth justice and balance. Teaching and reading for increased socio-environmental knowledges and deepen knowledges of democracy is essential for the ecopedagogical output of widened praxis to the global and the planetary.

There are various models of teaching that have been associated with global and planetary connectedness within ecopedagogical tenets such as critical global citizenship and its education (critical GC/E) and planetary citizenship and its education (PC/E).[6] The critical and dialectic foundation of ecopedagogy is essential for people to authentically participate as a citizen of their communities, world, and Earth. These two will be discussed more in Chapter 6. It is essential to note here that GCE models form a contested terrain that parallel globalizations contested terrain of processes from above and from below, with the emphasis of critical GC/E models coinciding with ecopedagogical tenets (Misiaszek, 2018b, 2016a, 2015).

3.3 Countering banking education

Countering banking environmental education models through Freire's work is essential because such shallow, non-transformative environmental pedagogies can be more socio-environmentally and unsustainably harmful than if the teaching did not take place (Misiaszek, 2018b, 2015, 2011). If environmental pedagogues teach that societal structures are non-transformable, planetary sustainability is not helped by environmental pedagogies but rather hindered—with the realization that schooling is only one, but a powerful educational aspect, for individuals/societies. What makes banking environmental pedagogical models so devastating within formal education is that knowledges and accompanying ideological formations are often not questioned by students, especially outside of higher education. This unproblematized ideological formation of students is a grounding factor of banking education (Freire, 2000; Apple, 2004). With teachers seen as sole knowledge holders within banking learning spaces, students have the uphill task of unlearning embedded ideologies taught as unlearnable because the unlearning task is taught as foolish and "abnormal." Returning to Freire's quote on the normalization of global neoliberalism in the last section and throughout this book, critical pedagogues must teach to counter rather than accept such economic structure and underlying ideologies.

Research on how environmental pedagogies and pedagogies on the environment construct ways of knowing and systematically/structurally influencing students' environmental actions, both positive and negative actions, is essential ecopedagogical work. This includes ecopedagogical research, but also within ecopedagogical reading and in ecopedagogical learning spaces. Environmental pedagogies that are determined to be banking environmental pedagogies, where "knowledge is a gift bestowed by those who consider themselves knowledgeable upon those whom they consider to know nothing" (Freire, 2000), are inherently the ones that must be questioned most on the overall environmental effects from such education models. One of the main reasons I found in my previous research on why such environmental banking education is more harmful than even absence of education is because schooling is usually unquestioned knowledge, so shallow environmental knowledges and little to no problematizing on environmental issues usually have more residue throughout the student's lives (2018b, 2015, 2011). Overall, instilling shallow and fatalistic ideologies of environmentalism, including possible ideologies of neoliberalism, from the "legitimate" authority of schooling counters all ecopedagogical characteristics and goals of praxis discussed throughout this book.

Environmental knowledges and experiences are devalued with only "legitimate" political knowledge being heard, such as neoliberal framing of all socio-environmental issues and possible solutions (Misiaszek and Torres, 2019). In banking spaces, "official" knowledges are never challenged—whether oppressive, empowering, (un)sustaining—silencing all voices but the teacher's(s'). It is impossible for socio-environmental transformation to emerge from such spaces. This is with the recognition, from Apple (2004), that this does not mean that students necessarily will not act toward ending oppressions through praxis; however, banking education, at best, does not support and, at worst, counters such praxis from students, leaning that schooling is an important factor of learning and ideological construction for students, but it is not the only one by far and many students rebel in various ways against their schooling.

> What is essential is that learners, though subjected to the praxis of the "banking system," maintain alive the flame of resistance that sharpens their curiosity and stimulates their capacity for risk, for adventure, so as to immunize themselves against the banking system.

Freire discussed above how students must maintain "flame[s] of resistance" when their schooling is banking education—an unfortunate relationship that students must have with their teachers (Freire, 1998b).

As banking environmental teachers "project an absolute ignorance onto others," not only is critical problematizing of environmental violence absent within the learning spaces but such problematizing is also not learned outside the spaces for students' critical literacy—to "read the world" and to "read Earth." Returning once again to the Gadotti and Torres' (2009) article on development and Freire in Brazil, they argued that years of banking education has largely contributed to the nation's lack of deep democratic participation in their own development. As the banking environmental educator "regards men as adaptable, manageable beings" (Freire, 2000), such teaching's residue continues outside of schooling making students "adaptable" to the environmental harms placed on them and legitimize such adaption from and onto others. Thus, taking on the socio-environmental oppression becomes expected and not countered, especially as democratic senses are systematically distinguished. In addition, banking environmental educators "minimize or annul the student's creative power" (Freire, 2000) without dialogue, so that praxis within learning spaces is nil, necessary problem-posing for socio-environmental transformation is absent, and thus being "environmental" is reduced to a superficial change, if any. Without teaching for creativity and innovation, possible world-Earth solutions become seemingly impossible and sustain/intensify fatalism. As discussed previously, ecopedagogy is impossible without utopic education because the politics of environmental violence will then be seen as insurmountable (i.e., Freirean-termed "limit situations" as insurmountable)—fatalism within the classroom translates to fatalism toward socio-environmental injustice and unsustainability.

Freire (2000, 1998b) discussed banking education's non-transformability through objectification of students and mystifying knowledges in teaching the labeled "unknowledgeable" students. Pedagogies that are fatalistic and non-dialogical mystify structures of oppression, including those from environmental violence, adhering to false "realit[ies]," as Freire discussed, in which humanistic, liberating teaching cannot occur.

> Domination is itself objectively divisive. It maintains the oppressed in a position of "adhesion" to a reality which seems all powerful and overwhelming, and then alienates by presenting mysterious forces to explain this power. (Freire, 2000)

Dominating entities rely on mystifying their constructed "realities" with neutral education that does not allow students to fully understand or question hegemony. Within such oppressive pedagogies, "power" is unexplained and mystified to be thought of within certain ideology(ies) determined by the "powerful." Education that acts as a tool to reproduce inequalities of power cannot act as a tool for socio-environmental transformation, with environmental devastation continuing without limits within the world (with there being, of course, limits of Earth's non-anthropocentric sphere according to the laws of Nature).

Regarding students as non-transformable agents of the world, according to Freire, banking educators view their students as "in the world," with the "assumption of a dichotomy between human beings and the world: a person is merely in the world, not with the world or with others; the individual is spectator, not re-creator" (Freire, 2000). This can be ecopedagogically widened in that banking education students are also non-transformable agents of Earth and "in the Earth" ("Earth" is purposely objectified with the article "the" here). This aspect of banking education is closely related to the other factors discussed, especially with teaching to instill fatalism. Extending the outsided*ness* of the student from the world, which further counters the student as connected with the rest of Earth—not as part of Nature but rather within the world as an object to follow acts of environmental violence without protest. This absence of protesting is twofold, as a person taught this way cannot be an agent of transformation, but also shallow banking education does not teach for deepened and widened world-Earth understandings that indict what and why protest in the first place. In other words, the knowledge of the root causes of unsustainable environmental violence is not taught, as well as ecopedagogical literacy as tools to "read" the politics of such violence. Lastly, socio-environment knowledge that is taught by "mythicizing reality, to conceal certain facts which explain the way human beings exist in the world" (Freire, 2000) does not allow students to use knowledges taught for widened-deepened understandings beside what the teacher(s) tell students, or for ecopedagogical praxis to emerge. The politics of the world with the objective knowledges of the non-anthropocentric sphere (i.e., laws of Nature), which are necessary for urgent, drastic actions that can only emerge from praxis, is systematically mystified through banking teaching. Such transformation is impossible without the ecopedagogical tools to determine the political roots causes of socio-environmental violence and planetary unsustainability which are purposely hidden through mysticism. All this strengthens the tool of banking environment for systematic *a*politicalization (or neutralization) of environmental violence.

3.4 Dialogue: Authentic and diverse

> Cooperation, as a characteristic of dialogical action—which occurs only among Subjects (who may, however, have diverse levels of functions and thus of responsibility)—can only be achieved through communication. Dialogue, as essential communication, must underlie any cooperation. In the theory of dialogical action, there is no place for conquering the people on behalf of the revolutionary cause, but only for gaining their adherence.... Authentic adherence is the free coincidence of choices; it cannot occur apart from communication among people, mediated by reality. (Freire, 2000)

In banking-education models students are defined as objects without any valuable knowledge, and are taught to absorb valuable knowledge from the only one(s) who have it—the teacher(s) (Freire, 2000). Dialogue is a cornerstone of Freirean Pedagogies, with him stating that "dialogue with the people is neither a concession nor a gift, much less a tactic to be used for domination" (Freire, 2000). Dialogue is a tool of social connectivity with one's own reality and gives the ability to understand others' realities. Banking education isolates students from their own realities, others' realities, and better understandings of their realities through others' perspectives of realities. Regarding denying freedom through banking education's isolating characteristics, Freire states:

> Education as the practice of freedom—as opposed to education as the practice of domination—denies that man is abstract, isolated, independent and unattached to the world; it also denies that the world exists as a reality apart from people. (Freire, 2000)

Beyond the social injustices placed upon the individual isolation, such banking techniques of separation denies self- and social constructions of what is "development" to determine actions through praxis. Reality and the "progress" of that reality (i.e., development) is denied, with development ideologies spoon-fed to students without any dialogue—without any democratic construction, without any self-reflection.

Without ecopedagogical dialogue, the (dis)connections with the rest of Earth are statically "told" to students without discussion for understandings of the non-anthropocentric sphere to fall in line with Development. Freire's arguments below on essentially reading the world can be reinvented to reading Earth with others through dialogue, as aligning more with what he wrote in his later writings (Misiaszek and Torres, 2019).

> Education as the practice of freedom—as opposed to education as the practice of domination—denies that man is abstract, isolated, independent and unattached to the world; it also denies that the world exists as a reality apart from people. (Freire, 2000)

As only the world is attached within banking model, the detachment of the rest of Earth is even more so as one to be commodified, for example, within neoliberal ideologies.

The non-anthropocentric sphere is increasingly distanced from the world by banking education's delegitimization of all other epistemologies, including ways of knowing that de-distances students' connection between them the rest of the world and Earth. Without dialogue that challenges epistemologies of the North, and their delegitimization of epistemologies of the South, Development ideologies will be reproduced through schooling without resistance. Severing authentic dialogue on socio-environmental issues only allows for the politics of environmental issues to sustain and intensify, with banking education as a Petri dish for nonresistance.

Dialogue within learning spaces is for better understanding through the true listening to others' knowledges and perspectives within the space, as well as discussions on learning resources brought into the learning spaces which provide other knowledges and perspectives (e.g., authors' perspectives written in their books, "official knowledges" in textbooks). Freirean dialogue is not a debate to win, but to listen and respond together to further deepen and widen understandings, with the realization that conflict is not negative but provides other aspects to understand one another and one's own perspectives and knowledges. Within ecopedagogical spaces, dialogue is for deepened and widened understandings of socio-environmental and sustainability issues from multiple knowledges and perspectives.

Ecopedagogical praxis is not a shift of power so the oppressed become the oppressors, but rather for global socio-environmental liberation—an important aspect in dialogue to problematize. Such dialogue problematizes roles of just power versus hegemony and also what is "success"—once again returning to Freire arguments on needing to radically change the concept of "success" as increasing the number of people "beneath" the Self. Freire also pointed out that often the oppressed that "succeed" according to this concept become worse oppressors than their previous oppressors. This, of course, is largely due to the non-dialogic ideological banking education that they received.

> [The] object of dialogical-libertarian action is not to "dislodge" the oppressed from a mythological reality in order to "bind" them to another reality. On the contrary, the object of dialogical action is to make it possible for the oppressed, by perceiving their adhesion, to opt to transform an unjust reality. (Freire, 2000)

Within ecopedagogical spaces, the utopic reality discussed includes liberation of the world holistically and planetary sustainability. Ecopedagogical dialogue should be full of questioning and re-questioning what the socio-environmental utopias are for those inside and outside the learning space, as it connects with possibilities of globally ending oppressions and Earth's unsustainability. These are the goals of ecopedagogies rather than a shifting of the source of oppressions and domination, as well as who benefits and who suffers. Such dialogue will be further discussed later in this chapter on Freire's concepts of generative themes emergent from limit situations, and utopic education.

Difficulties and, oftentimes, frustrations will surely emerge from teaching through dialogue, which will unquestionably be learning spaces of conflict when everyone is able to express their own thoughts, knowledges, and experiences, which is part of the process toward authentic dialogue. Freire has expressed in various ways that conflict is what makes us human, in that we all have our own understandings of the world, our own thematic universes that are "the reality which mediates men, and to the perception of that reality held by educators and people, . . . to find the program content of education" (Freire, 2000). As humans are social beings (Scherr, 2005), these are not individually constructed inside self-contained vacuums, but rather constructed socially with others within distinctions of one's Self. Social constructions of self-individuality largely counter Western concepts of strong individualism as paramount. Conflict frequently arises when one is told they are not 100 percent their "own person." In particular, social constructions of "individuality" strongly counter neoliberal ideologies of valuing one's own Self and that their individual*ness* and social positionality was due to their own actions and intellect.

It is the diversity of knowledges and perspectives, along with the conflicts that arise, that deepens and widens ecopedagogical spaces. This is opposite to banking education spaces that value homogeneousness in understandings and learning outcomes within neutral spaces that are absent of conflict. In other words, conflict is essential to Freirean and all critical pedagogies but, as Apple (2014, 2006, 2004) has expressed, conflict is most frequently structurally avoided in schooling so as to not question teachers and grounded ideologies. Rather than having knowledges and ways of thinking confined to a single set of ideologies predetermined from a static curriculum structuring specific ideological learning outputs (i.e., banking-education objectives), teachers and students construct together how the various "course" topics will be discussed—the "curricula."[7]

When I wrote in the previous paragraph that dialogue includes "discussions on learning resources brought into the learning spaces" I was essentially referencing curricula. Curricula is fluid and diverse and depends on who is defining it. Curricula which I am currently speaking of are the resources and ways in which topics are discussed, both within a predetermined "curriculum" from educational structures (within formal schooling). It is <u>not</u> the amount of didactic information learned in the curricula that measures the quality of ecopedagogies, but the depth and width of understanding new and prior knowledges. This counters many pedagogical models, including banking environmental ones, in that teaching environmental issues in which it is not the amount of didactic environmental information that makes good teaching but learning spaces that allow for diverse dialogue about socio-environmental oppression, unsustainability, and necessary transformational actions.

An important aspect of curricula Apple (2004) has argued is that problematizing the curriculum of what it does not contain is just as important as what it includes. What ideologies, from whose perspectives, and from what sources are present or absent in the curricula? For example, when past environmental disasters are taught, whose perspectives are voiced in the teaching resources? Whose perspectives in the learning space have a voice and which ones are silenced, such as pedagogically silencing students' voices within banking models where only the teacher(s) has a voice and official curricula could be viewed as having a metaphorical voice. Such silencing is inherently oppressive through human-domesticating education, as Freire expressed below.

> [Humans] become domesticated by the new contents which the leaders deposit in them. In other circumstances, they may become frightened by a "word" which threatens the oppressor housed within them. (Freire, 2000)

A key ecopedagogical question is, how can environmental teaching problematize what perspectives are absent to then bring those silent voices into the learning spaces, with specific attention to voices that are most socio-environmentally oppressed? As Freire has stated in the quote, what are the aspects of authentic dialogue in speaking "a 'word'" that make it frightening in threatening the oppressor and why, with analysis of the politics of the word? With this question too is what education teaches the oppressed to be frightened to question, or as "threatening" the oppressor? Problem-posing dialogue of whose voices are missing and how do we get those voices into the ecopedagogical space is essential—as grounded within critical pedagogies (Apple et al., 2009), the "listening" to of voices from those who struggle the most is central to all critical pedagogies. Without needing to be stated, what must also be problematized is, what are the politics of such silencing?

Also included in ecopedagogical dialogue is problem-posing what epistemologies we are learning through, as de Sousa Santos (2018) discussed Freirean pedagogies in his following quote.

> Freire, inspired by liberation theology and Marxism, proposed a paradigmatic shift: turning education (beginning with adult literacy) into a process of consciousness raising . . . by encouraging the production and acquisition of relevant knowledge in order to identify critically the concrete conditions of life and change them by means of a politics of liberation. Freire's consciousness raising is no idealist concept turning society into the consciousness of itself. Quite the opposite, it means getting to understand social structures as modes of domination and violence, and freedom as the refusal to accept such structures as destiny.

Authentic dialogue is an essential learning tool for understanding beyond one's own epistemologies, experiences, and knowledges, especially for those who are sociohistorically silenced. As will be discussed later, bringing in theories and multiple disciplines to help better understand others' voices outside of those in a learning space is also necessary, with the reflectivity of the limitations of knowing others' positionalities and epistemological lenses. Here it is important to expand a bit on limitations of knowing the world through self-reflection, in which humility is seeking what are the limitations in opposition to hubris that lead to increased ignorance with beliefs that our self-reflection has no limitations. De

Sousa Santos (2018) utilized the scholarship of Bourdieu to explain the need for self-reflexivity beyond one's Self.

> [According to] Bourdieu, a demanding exercise in self-reflexivity cannot but strengthen the belief in the monoculture of valid knowledge propounded by the epistemologies of the North. There is no room for bringing into account other ways of knowing that might correct or overcome the past failures of previous scientific knowledge or that might deal with other sets of issues.

Beyond one's Self is especially true when her or his schooling and outside education is taught within a single epistemology (or non-diverse set of epistemologies), and even more so if within epistemologies of the North. Such epistemologically limited self-reflection only strengthens beliefs of monoculture validity of previously held ways of knowing. Dialectic problem-posing not only questions what knowledge systems are used to understand the world-Earth; but one's self-reflection must also have the understanding that "the inquiry about limits is not an inquiry without limits" (de Sousa Santos, 2018). Self-reflection must be dialectic through ecologies of knowledges with the recognition of what is not known of the world-Earth and limitations of perceiving the world-Earth through others' ways of knowing (or could be more popularly termed as "through others' eyes"). This is with the previously discussed aspects of needing to problematize the limits of knowing the non-anthropocentric sphere and the laws of Nature all inclusively for the Self but also the world overall. Dialect teaching through ecologies of knowledges can deepen and broaden our understanding of the non-anthropocentric sphere and Nature's laws, but understandings of our cognitive limitations are also needed.

Freire discussed the devastating aspects of silencing the othered by taking away their right to be heard which leads to taking away their abilities to share their histories—in effect, taking away their histories. Freire's (2000) quote follows the previous quote by him on domesticating education.

> There is no historical reality which is not human. There is no history without humankind, and no history for human beings; there is only history of humanity, made by people and (as Marx pointed out) in turn making them. It is when the majorities are denied their right to participate in history as Subjects that they become dominated and alienated.

This coincides with how Albert Memmi (1991b) discussed that one of colonizations' most oppressive actions was/is structurally taking away the histories of the colonized. Freire has expressed in the quote how teaching that structurally objectifies humans, in which taking away their histories is an act of objectification, also instills fatalism for their futures. As previously discussed, Freire argued that one's histories non-fatalistically construct their futures, in which devalued, absent histories influence coinciding future(s) of oppression. Ecopedagogical analysis would focus on ending teaching that reproduces of socio-environmental oppressions by denying participation through dialogue to those who have been sociohistorically oppressed by environmental violence. Taking away their "right to participate in history as Subjects" within learning spaces invalidates their suffering from environmental in/justice.

There are difficult situations which will most likely occur in ecopedagogical learning spaces that offer safe spaces for authentic dialogue,[8] which are when student(s) do not become "environmentalists" or become more "anti-environmentalist" through the teaching. As discussed earlier in this section, the respecting and valuing of diverse perspectives within Freirean learning spaces, including conflicting with environmental teachers, are likely to produce self-identifying "environmentalists." Critical/Freirean pedagogues do not teach students what to think, that would be banking models, but rather how to better critically understand the environmental issues as hand, from multiple perspectives.

> It is not our role to speak to the people about our own view of the world, nor to attempt to impose that view on them, but rather to dialogue with the people about their view and ours. (Freire, 2000)

Such deepening and widening characterizes dialogue that is diversely structured through diverse perspectives and knowledges, with specific concern for deconstructing how sociohistorical ways of knowing (i.e., epistemological framings) can be socio-environmentally empowering or oppressive/dominating (Misiaszek and Torres, 2019). I have often posed the following question to my students: If our classrooms are not able to accomplish such dialogue, how can we expect the much larger, diverse world to do so?

Students' and teachers' perspectives can either coincide with or be against "environmentalism," which does form daunting conundrums with the ecopedagogical goal of praxis for socio-environmental justice and planetary sustainability. Also, opposite to banking models that discuss its absence, politics of the teacher(s) is acknowledged within Freirean spaces, including their environmental politics. The word "acknowledged" is used because all learning spaces are political (á la Freire (2000, 1998c)), but it is whether false neutrality will be artificially implemented or not that matters (Freire, 2000; Apple, 2004).

These conundrums are not negative in the sense of education, but there are value and ethical aspects of justice and oppression that are difficult to overcome as critical pedagogues and ecopedagogues.

There are many arguments given by scholars on such students-teacher(s) conflict conundrums within dialogue, but I would like to discuss the aspect that such conflict leads also to problematizing false truth held by students or by teachers. False truths are taught to purposely lie that the world is significantly affecting the rest of Earth, with one of the most obvious examples of this being climate-change denial. Without safe ecopedagogical spaces for authentic dialogue, such false truths are likely not challenged. I would state such learning spaces are increasingly needed in current era post-truthism, in that avoiding such conflicts strengthens falsities by removing the politics and histories of the creation of post-truth-isms, better understanding the tensions between the framings within socio/environmental (im)balances. Challenges of constructing safe dialogic spaces with intensifying post-truthism is largely due to the absence of common environmental truths to base dialogue upon (aligning with Foucault's arguments) seeing that opinions have become "truths" (Misiaszek, 2020). It leads to not listening to others' perspectives with the internet allowing learners to search and find only environmental information/knowledges that agree with one's own preconstructed opinions and ideologies (Misiaszek, 2020). This will be discussed more later in the book, but it is important to note how current conditions, such as globalizations from above and/or post-truthism, increase the complexities, but also the need, for creating safe dialectic pedagogical spaces.

Transformative praxis relies on authentic dialogue, as Freire expressed in the quote below.

> Cooperation leads dialogical Subjects to focus their attention on the reality which mediates them and which-posed as a problem-challenges them. The response to that challenge is the action of dialogical Subjects upon reality in order to transform it. Let me reemphasize that posing reality as a problem does not mean sloganizing: it means critical analysis of a problematic reality. (Freire, 2000)

Dialogic teaching provides human collective subjectivity to end environmental violence that lead to their oppressions, as well as others' and devastation within the non-anthropocentric sphere. In a book on ecopedagogy and GCE, I (2018b) used Freire's words from the above quote on needing "authentic adherence" for revolutionary transformation, as it "is the free coincidence of choices; it cannot occur apart from communication among people, mediated by reality"

(Freire, 2000). Authentic dialogue allows for democratic decision-making to better understand diverse realities of socio-environmental oppressions and unsustainability to cooperate with one another to determine what actions must be taken, and then to act. Freire's wording emphasizes the need for critical analysis of a problematic reality, rather than shallow sloganizing. Freire's discussions on leadership and revolution discussed in Chapter 4 of *Pedagogy of the Oppressed* (2000) very much coincides with the need of teaching for actions toward socio-environmental revolution grounded in planetary sustainability.

Some might ask if such politics should be within learning spaces, especially within formal schooling. As discussed previously, Freire argued throughout all his work that apolitical education is impossible. There are various articles and book chapters that Freire wrote in the title of education as cultural action (Freire, 1998a, 1985, 1971, 1970) and a compiled book entitled *Teachers as Cultural Workers: Letters to Those Who Dare Teach* (Freire, 1998d). There are many passages that could exemplify his work, with the following connected directly to dialogue.

> The teacher-of-the-students and the students-of-the-teacher cease to exist and a new term emerges: teacher-student with students-teachers. The teacher is no longer merely the-one-who-teaches, but one who is himself taught in dialogue with the students, who in turn while being taught also teach. (Freire, 2000)

Once again it is not telling students what to think but learning and teaching with everyone in the space toward deepening and widening understanding of others and their socio-environmental perspectives.

In the talking book *We Make the Road by Walking* (1990), in the form of dialogue between Freire and Miles Horton, Horton argued the sometimes nondemocratic dialectic aspect of calling a teacher a "teacher."

> Freire: . . . instead of naming a school for adults, I named the space and the students and the teacher "Circle of Culture" in order to avoid a name that sounds to me too much like traditional school. Instead of calling the teacher "teacher," I named him or her "coordinator of discussion, of debate, dialogue." And the students I called "participants of discussion."

Later Horton talked about how his Highlander Education and Research Center was not an organization's training school but an educational institute.

> Horton: The public who only saw that didn't know what went on at Highlander, and they assumed that we were an organizer's training school. But I kept saying no, no. We do education and they become organized. They become officials.

> They become whatever they are, educational directors. Basically it's not technical training. We're not in the technical business. We emphasize ways you analyze and perform and relate to people, but that's what I call education, not organizing.

Both of Horton's quotes reemphasize previous discussions in this section on conflicts of the politics of environmental educators teaching to produce environmentalists. Organizing is "pushing" an ideology(ies) for direct action as compared to critical education where it is the students determining on their own understandings and ideological formations. As subjective and social beings, dialectic pedagogies have teachers and students learn together to analyze their societies, and the world and Earth overall with the endless complexities between these local-to-global to planetary spheres.

In Freire's naming the spaces "Circle[s] of Culture" and Horton's emphasizes differences between training and education within making and "becoming" organizers, both emphasize the democratic and freedom aspects of passive, pragmatic training versus dialectic teaching. Unlike environmental justice organizing and its training, ecopedagogues' desired endpoints are not so pragmatic to environmental actions but student becoming environmentalists by self-reflection through dialogue that leads to praxis. Humans' unique ability of dialogue between one another allows them to have praxis of theories and reflection, as Freire expressed in the quote below.

> Animals, which do not labor, live in a setting which they cannot transcend. Hence, each animal species lives in the context appropriate to it, and these contexts, while open to humans, cannot communicate among themselves. But human activity consists of action and reflection: it is praxis; it is transformation of the world. And as praxis, it requires theory to illuminate it. Human activity is theory and practice; it is reflection and action. It cannot . . . be reduced to either verbalism or activism. (Freire, 2000)

Praxis, as expressed here and as focused upon in the next section, should emerge from ecopedagogical work to guide development, including our actions for development. As another example, some scholars point to as Freire being "antienvironmental," his argument isn't about (de)valuing humans from other animals but rather our unique communications ability for praxis. Freire also isn't saying that animals don't communicate, but rather historical, theoretical reflection is beyond "either verbalism or activism." This quote also reiterates what the aspect that only humans can develop through transformation, either positively or negatively, while animals can only adapt and evolve, and react to stimuli. This reflectivity, hopefully within praxis, gives humans the responsibility of the sustaining of Earth, as it is choice of conscious (un)sustainability. With this statement, the ecopedagogical responsibility is teaching through dialogue, through theoretical lenses, and with democracy for praxis of socio-environmental justice and planetary sustainability.

In addition, it is through dialogue with one another that praxis emerges—as Freire (2000) discussed below again that praxis does not emerge from isolation, but rather within spaces of critical dialogue.

> Thinking that is concerned about reality, does not take place in ivory tower isolation, but only in communication. If it is true that thought has meaning only when generated by action upon the world, the subordination of students to teachers becomes impossible.

This does not mean that praxis only emerges from dialectic spaces, but rather that it is constructed through histories of dialogue to construct social and individual reflection. Once again this raises the question of what we bring into dialogical spaces, including the curricula, for effective praxis. Whose perspectives, whose epistemologies, and whose overall thematic universes will be drawn from for learning, or not learning, about socio-environmental violence/peace and planetary violence for reflectivity of students' and our actions? Whose voices will be "heard" and "read," and whose won't be in learning spaces?

3.5 Praxis-based: Transformative

> The notion of "praxis" in traditional Western philosophy is more analytically complex than is the notion of practice or conflict per se. Indeed it is more appealing for those who would like to consider the relationships between theory and practice as blended together in the notion of praxis, a dialectical concept. (Gadotti, 1996)

There are many definitions of praxis, but I will utilize Gadotti's definition of praxis above as actions through dialogue and theories. With this simple definition that Gadotti began his book *Pedagogy of Praxis: A Dialectical Philosophy of Education*, I will further expand on ecopedagogical praxis, which has been briefly discussed already. Freire has discussed the inseparable connections between education and society as influenced by "development," and as tools for "development," which helps construct "development" ("development" substituting the phrase "overall plan for the society").

> When we are dealing with social practice in which the struggle for production, class conflict, and creative action are all dynamically interrelated, we discover that education is a process that takes social practice as the basis for learning

> and study. Education is itself a dimension of social practice. It seeks to know the reason for the practice and, through this knowledge as it deepens and develops, it also seeks new practice that is consistent with the overall plan for the society. (Freire, 1978)

The ecopedagogical goal of deepening and widening critical, dialectic reflection within and between world-Earth spheres is for determining what actions need to be done, such as how to reconstruct development. Within technocratic language this could be termed as the desired outcome(s) or resultant(s); however, the pathway toward teaching this should be opposite of being technocratic and that goals are plural rather than fixed and singular.

Toward this goal of *d*evelopment is the "move[ment]" that Freire discusses as from a self-consciousness of *being oppressed* to the consciousness *of the oppressed*, in which ecopedagogical teaching and analysis widen this to center on planetary oppressions as well.

> Men who are bound to Nature and to the oppressor in this way must come to discern themselves as persons prevented from being. And discovering themselves means in the first instance discovering themselves as Pedro, as Antonio, or as Josefa. This discovery implies a different perception of the meaning of designations: the words "world," "men," "culture," "tree," "work," "animal," reassume their true significance. The peasants now see themselves as transformers of reality (previously a mysterious entity) through their creative labor. They discover that—as people—they can no longer continue to be "things" possessed by others; and they can move from consciousness of themselves as oppressed individuals to the consciousness of an oppressed class. (Freire, 2000)

Within certain ecopedagogies, including the models that I discuss within this book, there is reinventing of Freire's interpretations of "world," "men," "culture," "tree," "work," and "animal"; however, it is the reconstructions of the perceptions of terms which Freire is focusing on most.[9] In other words, the problem-posing of these and other terms as we discuss labor as development, with critical questioning of whose *d*evelopment or *D*evelopment, is the essence of Freire's quote and can be deconstructed further within terms of ES*D* or ES*d*. How do these terms coincide with consciousness and with labor as one's and/or another's development, within the larger aspects of the rest of Nature, both within Freire's later writings (some of which has already been given throughout this chapter) and outside of them.

With some anti-Freirean scholars taking quotes like the one above to demonize Freire as antienvironmental, removing historical positioning of his writings, and that he was discussing the uniqueness of humans as reflective, sociohistorical beings capable of praxis to devalue all else of Earth. This would be similar to convicting Dewey (1963) as an anti-environmentalist because he discussed that humans shouldn't be trained as horses and dogs because of our self-reflective capabilities. Freire (2004) discussed this human uniqueness utilizing Marx's *Capital* (1977) with the example of bees compared to architects.

> At a certain point in Capital, while discussing human work as opposed to that of other animals, Marx says that a bee could not possibly compare to even the most "modest" of architects. After all, a human being has the capacity for ideating an object before ever producing it. The carpenter has the table drawn up in his or her head before building it. This inventive capacity implies a communicative one, on all levels of the vital experience. The creative and communicative activities of human beings, however, connote qualities that are exclusively their own. Communication exists in life, but human communication is processed as well, all especially so, in existence, a human invention.

Freire here is not diminishing the work of bees or their needs within by using Marx's wording, but rather differentiating our ability as humans to "invent," and consequentially "reinvent" our world and Earth. Bees are in fact one of the most needed beings on Earth as they pollenate plants and flowers resulting in food, oxygen, and many other necessary aspects of the world-Earth. Unfortunately, there is a current environmental crisis due to massive death of bee populations worldwide. As I have stated previously many times, it is our responsibility as self-reflective (i.e., praxis), sociohistorical beings to maintain world-Earth sustainability, due to the fact that our actions are the reasons for imbalance leading toward violence and injustices. This also links to our work on labor, and its structures, toward (in)justice and *d*/*D*evelopment.

Determining actions through endless problematizing of what is development largely coincides with problematizing of what is "success," as Freire initiated such analysis in the first chapter of *Pedagogy of the Oppressed* (2000), is essential for ecopedagogical praxis. Goals of dominance over the non-anthropocentric sphere that coincides with *D*evelopment must be deconstructed if we are to achieve world-Earth sustainability, as they ultimately lead to socio-environmental oppressions and dominance. This includes unlearning many knowledges emergent from dominant epistemologies, especially those which de Sousa Santos (2018, 2016) names epistemologies of the North. A key one would be literature/imagery promoting the controlling of Nature as a sign of "success" rather than literature/imagery for living *with* the rest of Nature. The latter counters the former which are long-held ideologies, especially from the West, of power and supremacy of humanness. False critiques of Freire see his work as the former, but I, with Carlos Torres, very strongly argue that his work is the latter (Misiaszek and Torres, 2019).

Praxis for world-Earth peace and sustainability must deconstruct development from being directly related to the controlling of the non-anthropocentric sphere, to development that aligns with this space within the laws of Nature—coinciding very closely to Illich's (1983) arguments previously discussed. Freire argued how extremely difficult it is to critically challenge one's own ingrained ideologies even when the ideologies form much of their oppression. The challenging of long-held ideologies in ecopedagogical spaces is also difficult with the added aspect that the non-anthropocentric sphere cannot be fully known (i.e., limits of knowing Nature's laws). Freire (2000, 1998b) discussed how the pursuit of freedom is, in part, overcoming the fear of becoming something unknown to the Self and that freedom, or "success," should not be becoming the oppressor. Ecopedagogical work, and guiding ecopedagogical praxis, is overcoming such fear and radically altering the dogma of "success" in our interactions/connections with the non-anthropocentric sphere.

In the below passage, Freire discussed how the oppressed have oppressors within themselves who need to be self-countered.

> [Humans] become domesticated by the new contents which the leaders deposit in them. In other circumstances, they may become frightened by a "word" which threatens the oppressor housed within them. (Freire, 2000)

The "oppressor housed within them" is not one of pragmatic domination but rather how the oppressed see themselves through the eyes of the oppressor by internalizing their subaltern*ness*. Nor is it binary, in which there are those who are the oppressed and those who are the oppressor but rather as everyone is both the oppressor and the oppressed. Domestication, as Freire wrote in the quote, needs continuous ecopedagogical problem-posing of how *D*evelopment usage of the non-anthropocentric sphere is justified by the neutralization of its politics to be false, as it is falsely portrayed as *d*evelopment. In other words, it is the domestication of the oppressed by systematically silencing voices of misuse of the rest of Earth and thus stopping "uncomfortable" socio-environmental dialogue necessary for transformation and ecopedagogical praxis (Misiaszek and Torres, 2019). The fear is countering *D*evelopment which "the oppressor" within the oppressed has been taught as unquestionable "progress," or as measurements of "success" as Freire (2000) vehemently argued against.

There are various aspects of "success" that must be challenged, especially within normative ideologies of neoliberalism and livelihood, as discussed previously. As a key example, material consumption as a measurement of individual success and societal development without bounds are ideologies that cannot lead to planetary sustainability or socio-environmental justice, thus ecopedagogical praxis would inherently be grounded in countering them. Reconstructing our goals within the relative, non-binarily fixed oppressor-oppressed continuum must also be within/between the widened world-Earth spheres that challenge notions of Development goals. As discussed throughout this book and more intently in Chapter 5, everyone is negatively affected by unsustainable environmental violence but there is a wide range of degrees of socio-environmental suffering. Ecopedagogical praxis is for utopic goals of globally wide socio-environmental peace and sustainability through critical problematizing the inequalities of socio-environmental inequalities. In other words, how can we reconstruct "development" and "success" for liberation for the world holistically and sustainability with the non-anthropocentric sphere for development rather than shifting oppressions from Development to others?

A key reason why ecopedagogical learning for self-defined praxis is so important is because praxis is also what makes us human. Only humans are able to construct praxis, and teaching that does not allow it within learning spaces and/or ideologically discourages it outside of them is inherently dehumanizing, as Freire (2000) expressed below.

> The difference between animals—who (because their activity does not constitute limit-acts) cannot create products detached from themselves—and humankind—who through their action upon the world create the realm of culture and history—is that only the latter are beings of the praxis. Only human beings are praxis—the praxis which, as the reflection and action which truly transform reality, is the source of knowledge and creation. Animal activity, which occurs without a praxis, is not creative; peoples transforming activity is.

We are the only beings that can understand the world and Earth through our and others' histories, with realization of the limits of our understandings. These aspects have been discussed previously but, in this section, it is in terms of praxis. In other words, how do we reflect upon our actions? What do we reflect upon? Whose perspectives do we reflect upon when our determining actions? Do we consider the non-anthropocentric sphere on our actions?

These questions can be very difficult to self-reflect upon because, as ecopedagogues and environmentalists, some of our actions can be viewed as hypocritical. Such as, do we eat meat, do we take hot and/or too-long showers, and do we travel long distances that require plane transportation (which is one of the most environmentally harmful ways to travel)? Without needing to be stated, I have found myself being a hypocrite within these examples, and many more which question my own praxis.

Similar to this line of reasoning, I return to the Freirean tenets of safe spaces for authentic dialogue, conflict, and the conundrums of non-environmental dialogue and, as the focus here, non-ecopedagogical praxis from students. Creating safe spaces is difficult for teaching, but also needs constant self-reflection from ecopedagogues when students don't have the same goals of praxis as them. To expand on the example given previously, what if students who were climate-change deniers remain deniers after a quarter, a semester, or a year(s) of taking ecopedagogical courses? In my years of teaching a graduate-level ecopedagogy course, I have run into ideological conundrums between socio-environmental justice and sustainability. The complexities of differing goals of praxis is essential for continuous self-reflection with deepened and widened analysis of all these factors, but also student actions' through reflection will not necessarily coincide with ecopedagogical praxis.

Praxis for *d*evelopment rather than *D*evelopment is the goal of ecopedagogies. Emphasizing upon previous arguments, humans as beings capable of praxis lead to both social justice and injustice, and planetary sustainability and unsustainability.

> It is as transforming and creative beings that humans, in their permanent relations with reality, produce not only material goods—tangible objects—but also social institutions, ideas, and concepts. Through their continuing praxis, men and women simultaneously create history and become historical-social beings. Because—in contrast to animals—people can tri-dimensionalize time into the past, the present, and the future, their history, in function of their own creations, develops as a constant process of transformation within which epochal units materialize. (Freire, 2000)

I read this quote as Freire trying to emphasize "production" as beyond the material and beyond economic development, especially *D*evelopment, but also "social institutions, ideas, and concepts." This latter aspect is more important than the first; however, development is often only considered the former as material production with economic profit, and neoliberalism and globalizations from above only having *D*evelopment frameworks. Ecopedagogical work problem-poses how we "tri-dimensionalize time into the past, the present, and the future, their history" (Freire, 2000) to transform into the future—asking and re-asking how we develop from this tri-dimensionalizing. As discussed previously, Freire did not have a fatalistic concept of history dictating our future but what is essential is to critically problematize what histories are constructing futures. Futures' plurality emphasizes possible multiple futures that can emerge from histories and the present, with the realization that it is the tri-dimensionalizing to construct futures toward self/social constructed utopias. Banking education is inherently constructed to only have a single, static future in which it structurally reproduces unsustainable socio-environmental violence—as such, ecopedagogical praxis does not emerge from banking environmental education models.

3.5.1 Radical, revolutionary ecopedagogical praxis

Radical praxis emerges from democratic, dialectic education that is bottom-up, as opposed to top-down banking-education models that are absent of dialogue, especially that of possible revolutionary actions to end oppressions. In his last chapter of *Pedagogy of the Oppressed* (2000), Freire discussed revolutionary leaders and organizations as needing to be "in communion [to] liberate each other," as the two quotes below express.

> [We can] legitimately say that in the process of oppression someone oppresses someone else; we cannot say that in the process of revolution someone liberates someone else, nor yet that someone liberates himself, but rather that human beings in communion liberate each other.
>
> ... the oppressed and the leaders are equally the Subjects of revolutionary action, and reality serves as the medium for the transforming action of both groups. In this theory of action one cannot speak of an actor, nor simply of actors, but rather of actors in intercommunication.

Throughout his writings, Freire discussed how praxis must emerge from bottom-up, democratic organization/leadership that is full with solidarity, as his quotes expressed. The aspects of being "with" is not one of shallow wording but one of solidarity—"acting together in unshakable solidarity ... this solidarity is born only when the leaders witness to it by their humble, loving, and courageous encounter with the people" (Freire, 2000). Without such solidarity, authority becomes oppressive authoritarianism, as discussed previously.

Intensifying globalizations has greatly furthered the need to include the role of being an active member of one's own local society, as well as being a member (or, a global citizen) of the world. In Chapter 2, I discussed how globalization from above is distancing voices that lessen and weaken democracy; however, globalizations from below can provide possibilities to de-distance, and amplify local voices toward thickening democracy. Ecopedagogical praxis must be local-to-global in scope to determine what necessary radical transformation is essential and how globalizations from above can be countered to strengthen those from below. Praxis to determine such organization and leadership must have thick solidarity globally for multiple reasons with the most important one being that most environment violence is global in negative causes and effects. Environmental devastations, especially those nearing to their "tipping points," are in need of solutions within the global sphere. Praxis to solve climate change cannot happen only in select sub-global spheres without global actions also. For example, ending fossil fuel usage only in the Global South is not going to end rising ocean levels. There are various problems with such problem-solving, not least the issues of intensifying hegemony of the Global North within such a solution. Many, including myself, would argue that critical global citizenship is essential, with or without the term itself being invoked (Syed and Misiaszek, 2020), to determine what needs to be transformed and how. Once again, all this is with the recognition that often ecopedagogical praxis leads to the need for radical, revolutionary transformation.

One important note is that Freirean, ecopedagogical praxis is not devaluing cultures and traditions when transformation is spoken about. Freirean, critical pedagogy is for transformation through endless problematizing, but it is not to end local cultures toward a specific endpoint, especially a globally homogenous endpoint. This is what many anti-Freirean environmental pedagogues get wrong in that constant transformation means devaluing current cultures toward a single endpoint of an overly simplified "better world" (Au and Apple, 2007; McLaren, 2007). Freire wrote extensively on the negative aspects of cultural invasion, especially with intensifying globalization and neoliberalism (Freire, 2000, Freire, 1992, 1985, Freire, 1978), including the following quote.

> [The] object of dialogical-libertarian action is not to "dislodge" the oppressed from a mythological reality in order to "bind" them to another reality. On the contrary, the object of dialogical action is to make it possible for the oppressed, by perceiving their adhesion, to opt to transform an unjust reality (away from cultural invasion). (Freire, 2000)

Freirean problem-posing pedagogy is within local contexts toward a "better world" and problematizing that one person's/society's "better world" is not oppressive to others' concepts and realities of their "better worlds." Intensifying globalizations exponentially complicates the interconnections toward such problem-posing teaching. In addition, planetary concerns within ecopedagogical spaces exponentially complicate globality of local "better world" constructs to be with sustainable balance for the rest of Earth.

3.6 Bottom-up approaches

I have discussed bottom-up approaches throughout this chapter but in this section, I delve more deeply into some of the more pragmatic and theoretical aspects of this concept, including in the following subsection discussing Freire's use of Hegel's master-slave dialect for ending oppressions through bottom-up approaches. Ecopedagogies must be constructed with local realities which are meaningful to those in the learning spaces. In other words, education must emerge from the realities from what is outside the "classroom" (either tangibly or figuratively). This does not mean that what is learned about is only within the local surroundings, but it includes lessons learned through local meanings, knowledges, and realities of those in the learning space. Ecopedagogies include teaching outside of students' local sphere(s) to understand Earth, but how you teach "foreign" topics is through local epistemologies, histories, traditions, and all else that is meaningful to those in the learning space. Schooling, in this way, must be more entrenched *within* and *throughout* the local public spheres. The term "throughout" that I have used indicates Freire's (1978) arguments that schooling must be alongside all the societies/populations that they serve and not as an outside, elitist-intellectual entity imposing foreign knowledges and practices upon them without contextualization. Intellectual elitism systematically devalues all other previous knowledges to learn new knowledges *through*. This is not to say that outside knowledges should not be utilized within their communities for problem-solving, but that it is incorporated democratically *through* their previous knowledges, rather than forced upon them as mystified, superior knowledges that belittle all previous knowledges and practices.

Within ecopedagogies, learning environmental knowledges through previous local socio-environmental knowledges is essential. Teaching does include nonlocal knowledges and epistemologies that might not be present within more local understandings, such as widened understandings of larger planetary connections and deepened knowledges from outside the self-defined local spheres. It is through local understandings in bottom-up teaching that new understanding can be learned and, just as importantly, problematized.

> Critical pedagogical analysis, including ecopedagogical analysis of environmental pedagogies, would question the politics of why agroecology has been ignored, by asking who benefits from its absence and the legitimization of sciences. The dominant educational models as currently practiced do not seem designed to aid development for the masses. (HaLevi et al., 2018)

This quote appeared in an article that I wrote with various East African permaculture experts and critical pedagogues on the need for ecopedagogy within agriculture within the region for *f*arming, as opposed to *F*arming, with the same capitalization of *d*/*D*evelopment and coinciding with the lower-/uppercasing of "*f*/*F*arming."

Problematizing ideologies that are most often seen as oppressive is essential, but bottom-up reinvention must also locally contextualize theories, epistemologies, and ideologies that are viewed as countering oppressions, as Amílcar Cabral discussed.

> We needed to know them [Marx, Lenin, etc.], as I've said, in order to judge in what measure we could borrow from their experience to help our situation—but not necessarily to apply the ideology blindly just because it's a very good ideology. . . . Our desire to develop our country with social justice and power in the hands of the people is our ideological basis. ((Cabral, 1971) cited in (2018))

It is the local reinvention of outside theories that leads to true praxis—the use of praxis within the situations at hand. Although most environmental issues are within and between widened spheres, understandings emerge from local societies with action emerging from authentic dialogue with those from other societies. As discussed previously, Freire argued this within his concept of reinvention in which pedagogies and theories, including his own, were not to be blindly followed but reinvented to be meaningfully to the contexts of the current happenings at hand, while retaining the pedagogies' and theories' foundations. The ecopedagogical foundations of socio-environmental justice of the world and sustainability of Earth through Freirean scholarship is essential, but meaningful progressing toward ecopedagogical goals are taught, read, and researched within the contexts of local spaces. This is also essential for ecopedagogical praxis.

Bottom-up teaching is essential not only for socio-environmental justice for students but also for effective praxis for *d*evelopment, planetary sustainability, global socio-environmental peace, and other goals of ecopedagogies. Without local voices, knowledges, and epistemological framings, necessary socio-environmental transformation is impossible because outside knowledges are mystified within local spaces and those outside the local spaces cannot fully understand (or metaphorically "hear") socio-environmental oppressions upon the local societies. The latter part of this statement will be discussed further in Chapter 5, but here are some of Freire's (2000) key aspects on why bottom-up teaching is essential and bottom-up praxis is needed for those within the contexts of socio-environmental oppressions, but also for those outside to deepen their understandings of the oppressions for more effective praxis.

- Who are better prepared than the oppressed to understand the terrible significance of an oppressive society?
- Who suffer the effects of oppression more than the oppressed?
- Who can better understand the necessity of liberation? (Freire, 2000)

It is only when local voices are heard and local epistemologies are used that others outside of "the local" can truly be agents of necessary socio-environmental transformation.

De Sousa Santos (2018) and others have discussed how self-reflection when confined to one's own epistemologies, especially those grounded within sociohistorical oppressions (i.e., epistemologies of the North and sociologies of absences) only helps to reproduce rather than counter socio-environmental oppressions. In short, understanding and theorizing oppressions must be through those who suffer the most, with possibilities of utilizing outside knowledges and epistemologies through localized understandings of socio-environmental injustice and planetary sustainability. This work inherently counters globalizations from above, including processes toward <u>D</u>evelopment rather than <u>d</u>evelopment, by amplifying voices of sociohistorically othered.

3.6.1 Master-slave dialectic: Consciousization of being the hosts

> How can the oppressed as divided, unauthentic beings, participate in developing the pedagogy of their liberation?
>
> [Freire's Answer:] Only as they discover themselves to be "hosts" of the oppressor can they contribute to the midwifery of their liberating pedagogy . . . pedagogy makes oppression and its causes objects of reflection by the oppressed, and from that reflection will come their necessary engagement in the struggle for their liberation. And in the struggle this pedagogy will be made and remade. (Freire, 2000)

Freire (2000) problematized the very difficult question on how bottom-up transformation can occur if the "bottom" is powerless. Gadotti (1996) argued through Hegel's master-slave dialect that recognition that the masters' power comes from their oppressive hierarchy over the slaves, which corresponds to Freire's argument of problematic "success" (i.e., of the masters) defined by how many people (i.e., slaves) are beneath them.[10] Reworded in ecopedagogical terms, it is through democratic, dialectic discussions of the politics socio-environmental oppressions for the master's power on the back of the slaves who suffer the most from them.

It is through critical problematizing of who struggles the most from environmental violence and determining who is most benefiting from them, in which understanding emerge as to how to end the oppressed (i.e., slave) socio-environmental oppressions to take away the power of the oppressor to remain an oppressor (i.e., removing the power of the master to remain a "master"). To give an example within ecopedagogical reading of development, it is the recognition that the oppressors' *D*evelopment can only occur through *d*e-development of the oppressed. A key concern within ecopedagogies is how teaching can help students to deconstruct environmental violence through Hegel's dialectic, as well as the problems that emerge within environmental pedagogies that hide Hegel's dialect connecting power to others oppression, especially within banking models. Returning once again to Gadotti and Torres' (2009) article on Freire and education for development within Brazil, I view that their arguments could be seen through Hegel's dialect, in that it is only when Brazilians fully realize that they are the hosts of (neo)colonializer's *D*evelopment that liberation will be possible. They claim that years of banking education, if schooling at all, has systematically made the Brazilian masses ill-prepared for authentic democratic participation and determining actions for their *d*evelopment (Gadotti and Torres, 2009). Their arguments provide a good illustration of the concern I have raised.

With Freirean consciousization that socio-environmental dominance (i.e., from the "masters") emerges only from socio-environmental oppressions placed upon others (i.e., upon the "slaves"), recognition of oppressions can be continuously reread and transformative praxis can emerge. As Freire stated in the quote given to introduce this subsection, "reflection will come their necessary engagement in the struggle for their liberation," the power the oppressed (i.e., slaves) have upon their oppressors' (i.e., the master's(s')) power is essential to understand (i.e., consciousization). Actions for *d*evelopment counter the oppressors' *D*evelopment, and thus deny, to some degree, the oppressors' power.

Another critical ecopedagogical question of Hegel's master-slave dialectic is its possible incorporation, or not, of the non-anthropocentric sphere and planetarily holistic aspects of sustainability. I will not fully go into answering this question but possibly the power dynamics of unsustainability could have some application, or more radical reinvention of the dialect, the master-slave relationships between humans and the rest of Nature. With only humans as historical, reflective beings, what is the positioning of the hosting that we have with the non-anthropocentric sphere? Who, if any, counters the hosting upon the non-anthropocentric sphere which oppresses the enslaved and dominates Earth outside of the anthropocentric sphere? Or are we hosts of the rest of Nature? These are complicated questions which most environmental pedagogies grounded in epistemologies of the North will most likely never be problematize in learning spaces and even more unlikely within banking spaces.

3.7 Generative themes

Many of the students I have taught, especially in China, have argued that Freirean teaching is not possible within the different contexts of Chinese cultures and histories, including pedagogical foundations. Another argument is that it is impossible within large populations, with China as an example to this argument. This quite popular argument among noncritical pedagogues, especially from those favoring increased standardization and high-stakes testing (and endless re-testing), gets eventually brought up in my ecopedagogy class, as well as all my classes. I usually initiate such discussions, with both teachers and students, with the topic of Freire's generative themes, limit situations, and persons' thematic universes—all of which I have discussed previously and some I will give additional detail to.

Constructing generative themes is democratically determined between teacher(s) <u>and</u> students as guiding themes to problematize for teaching through. Within the learning space, however large it might be, it is teaching and learning through generative themes that help make the course's content more meaningful to everyone in the learning space. This does not mean that every topic is going to highly resonate with every single person in the space, especially within very diverse learning spaces, but rather the democratic decision-making through dialogue helps to determine some key socio-environmental issues more likely to be meaningful to all the students. Determining the commonalities and differences between diverse ecopedagogical themes is essential within the learning for everyone, including better understanding everyone's own positionality(ies), understandings, and epistemological framings.

In previous research on contextually defining what "successful" ecopedagogies are (2012, 2011), environmental pedagogies were found to often fail because they systematically teach us-versus-them differences of socio-environmental violence rather than also including the commonalities between contexts that initially seem very dissimilar. As Freire (2000) argued, divisions help to weaken bottom-up resistance to oppressions; teaching such absolute othering is a technique for sustaining/intensifying dominance—including teaching for Development. Teaching contextual differences is important; however, it is equally important to learn the commonalities of socio-environmental issues especially with environmental issues as global and planetary in scope for achieving true sustainability, as well as praxis toward this end.

I will discuss very briefly how individual/population's context within one's education is essential for freedom and socio-environmental praxis. Freire used the term "thematic universe(s)" as describing the Self's own understandings of their realities in that "the complex of their 'generative themes'—inaugurates the dialogue of education as the practice of freedom" (Freire, 2000). Without teaching through how each student understands socio-environmental issues within "their generative themes," the learning is unmeaningful and dialogue within learning spaces is unauthentic. Dialogue is unauthentic because discussions are forcibly adapted to repeating the teacher's (and larger schooling system's) understandings without giving the students the chance to express their own understandings or challenge either the teacher or the system.[11] As discussed previously through Freire's arguments, dialogue and the possibilities of subsequent freedom is not "gifted" to students but is a right. Without the ability to authentically discuss one's own thematic universe, together with others' thematic universes, students' rights as humans are taken away by silencing them, as well as systematically limiting possibilities of their ecopedagogical praxis.

> Problem-posing method—dialogical par excellence—is constituted and organized by the students' view of the world, where their own generative themes are found. The content thus constantly expands and renews itself. The task of the dialogical teacher in an interdisciplinary team working on the thematic universe revealed by their investigation is to "re-present" that universe to the people from whom she or he first received it—and "re-present" it not as a lecture, but as a problem. (Freire, 2000)

Problem-posing environmental violence through generative themes, within learning spaces and within one's own reading (i.e., ecopedagogical literacy), is an act of praxis—giving way to possibilities of transformational praxis. This aspect is eloquently stated by Robert Rhoads and Carlos Torres (2006), in that "constructing generative themes builds an agenda for action to surmount 'limit situations.'" Limit situations are the barriers between what is actually happening and what is possible, both determined by teacher(s) and students within learning spaces. Within ecopedagogical literacy, the process is through critical self-reflection, with the reflection including the limitations on the Self's own reflection. Generative themes are described as the following by Rhoads and Torres (2006):

> Refer to sets of issues that reflect themes within a particular epoch (particular point in time) and thematic universe of their epoch (subjectivity in reading the world) within which brings together a particular set of ideas, values, beliefs, hopes, doubts, and challenges.

Within learning spaces, dialectic education means to "[call] forth a set of opposites of these issues in a dialectical fashion" with the recognition that conflict will emerge because we are human, as Freire has argued, in that we have different histories and have differing "set[s] of ideas, values, beliefs, hopes, doubts, and challenges." Rhoads and Torres' (2006) discussions on "their epoch" indicate the unfinishedness of humans/societies in that we can transform within and between humans/societies, as opposed to fatalistic education in which knowledges, ideologies, and fate is falsely taught as predestined and unchangeable.

Knowledge is historically fluid over time within and between generations, that construct our understandings of the world, as well as being contextual. What makes our own identities is through these knowledges. Individual "identities" is plural because a person has not one but multiple identities. The multiple identities that individuals have both coincide and conflict with one another, with globalizations widening this conflict (although it could be argued also that in some cases identities are also narrowed by decreasing countering of one's own ideologies, as this argument would coincide with the previous arguments of the heterogenous*izing* or heterogenous*izing* of individuals and societies overall). Oppressive education, including banking education, often teaches as though individuals making up population(s) have only a single identity within the entire population/society which problematizes what Albert Memmi (1991b) termed as "the plural" within colonialism. Memmi's argument was that the colonizers thought of the colonialized population as a single identity—that is, if you "know" one, you "know" them all. Such ideology dehumanizes a whole population down to a single identity, in which that identity is specifically constructed to *other* the population as a whole.

Dialogue through generative themes are not only to be within one's own thematic universe but also to hear from thematic universes from everyone within the learning space and try to understand others' thematic universes outside of the space. Isabel Bohorquez (2008) argued that teaching must "pose obstacles to people becoming something more than they are—understanding these themes and wrestling with them is a step toward a fuller and more meaningful life" (Rhoads and Torres, 2006). It is in the intersections between diverse thematic universes that deepened and widened perspectives emerge, including determining what understandings and epistemologies are not included within the learning space, through democratic decision-making on what to "bring into" the learning space. Critical literacy, including ecopedagogical literacy, involves deepened and widened self-reflection on what perspectives are needed. It involves the critical questioning and re-questioning of what is needed in the learning: "what is needed to know and how can we/I learn it?" Within environmental issues, the questioning is endless as world-Earth complexity and "messiness" is limitless.

Within ecopedagogical spaces, what is needed is the problem-posing the gaps of environmental ills (social connections and planetary framing) of how "we" want the world (within the realization of contextual plural utopias) and the gaps in societal systems/structures, including education, and devising actions to this end through praxis. The grounding of critical pedagogies is how diverse utopias connect or disconnect with others' utopias toward worldwide social justice. The grounding of ecopedagogies is the aspect of how these utopias connect and disconnect with planetary sustainability, even within the aspects or the previous statements of critical pedagogical goals within the world. For the latter part of the last sentence, an example is decision-making on the permissible degree of environmental violence due to human wants and needs while still maintaining the balance with Earth. For example, what happens to people's homes within a forest when a fire starts? Should the fire be put out even though the burning of the underbrush is Nature's ways of "fire management"? What are the generative themes around these topics through multiple, diverse thematic universes and epistemologies, through understanding Nature's laws as well as our own limitations of knowing them fully?

Deepening and widening students' self-reflection within critical, dialogic teaching spaces is to "locate the seat of their [(i.e., students')] decisions in themselves and in their relations with the world and others, [and to empower] people [to] overcome the situations which limit them" (Freire, 2000). Ecopedagogical problem-posing extends this quote to the planetary sphere to determine what the environmental issues/actions events that cause social struggle are and how education sustains, intensifies, and/or decreases this. It also questions, as discussed throughout the last two chapters, the politics of all this, as well as the (dis)connections with planetary sustainability outside of the anthropocentric sphere. Is it through local lenses that global understandings are learned/read within and outside the learning/self-space, including the non-anthropocentric sphere with the recognition of limitations of understanding this non-anthropocentric sphere? This recognition of limitations is also with what one knows about the world, outside of their thematic universe, while learning from all others in the spaces and with critical, democratic decision-making with regard to what needs to be learned next, and then next after that, to continue without end. In the quote that began this paragraph, "locate[ing]" refers to locating one's own role as well as others' roles in determining actions through praxis, which increases world-holistic socio-environmental justice and planetary sustainability. In other words, it means to problem-pose how we locate ourselves (and others, as well from the Self and "their" Selves) within environmental decisions made. Through such problem-posing, what/who we consider and not consider, as well as the depth and width of such considerations, also emerges. It is through critical-based ecopedagogical learning and literacy that we can further deepen and widen these understandings, while respecting the limits to one's own understandings.

Determining limit situations within ecopedagogical spaces is problem-posing the educational gaps that narrow learning and that need to be taught next (or "read" next in ecopedagogical literacy) for transformational praxis to occur. Banking environmental education models do not lead to such transformational praxis because they reproduce the oppressive benefiting of socio-environmental injustice and unsustainability by not allowing for such critical problem-posing. As previously stated on ecopedagogical work, the histories of socio-environmental oppressions and unsustainability must be problematized, for understanding both histories and how our histories form our present and futures, especially in constructing individuals' thematic universes. In Freire's words, it involves better understanding our "relief from the background, revealing their true Nature as concrete historical dimensions of a given reality" (Freire, 2000). This leads to the following questions. What are the limit situations within our histories—such as colonializations, racism, patriarchy, and other social oppressions of othering—that are deeply engrained ideologies that must be countered? Does this include how we perceive the rest of Earth and why? Do we see this as something that must be sustained or something that is "there" for "our"/"their" benefit without regard or limitations? If Nature is to be sustained, is it only for populations of humans or is there also reasoning beyond anthropocentric concerns, as Nature is beyond just for satisfying humans' needs and wants?

I would argue that limit situations should not be defined as Natures' laws themselves, because limit situations are defined, by Freire (2000), as obstacles to liberation. Here I am expressing that Nature does not oppress but rather "exists" without consciousness with the goal of balance, in which the world it situated within as part of Nature. Limitations of natural resources and the occurrences of natural disasters do negatively affect humans, but Nature's limits and degrees of disasters are defined within the anthropocentric sphere as we are part of Nature. This will be discussed later; however, here it is important to restate that the non-anthropocentric sphere does not oppress us but exists with all accompanying mechanisms and limitations within which we must live, as a part of Earth (i.e., Nature). This is important to problematize when defining world-Earth limit situations.

Notes

1 In the English language, the terms "live" and "exist" have assumed implications opposite to their etymological origins. As used here, "live" is the more basic term, implying only survival; "exist" implies a deeper involvement in the process of "becoming" Freire (2000).

2 This notion of utopian marginality that Freire appropriated by declaring himself a "utopian traveler" is founded on a model of spirituality that is itself utopian. It is utopian because, as poets have suggested in countless verses, utopia is that distant horizon, a horizon that one always wants to reach but can never approach. One takes two steps toward this horizon and it recedes another two paces. What, then, is the advantage of utopia, we might ask? As a rational and spiritual model guiding our desire, it helps us to keep moving. (Torres, 2014)

3 In reference to the quote, I want to reemphasize two previous discussions in this book. First is that Freire is not stating in an antienvironmental way that all which is not humans on Earth is devalued, but rather we are unique is our ability of reflection, being historic, and dreaming of possible futures (Misiaszek and Torres (2019)). Second, ecopedagogues focus on deconstructing and reinventing development with one another within the global and sub-global spheres with the laws and valuing of the non-anthropocentric sphere planetary (i.e., holistically in the planetary sphere).

4 This is worded as "temporary" to reflect that Earth will exist and return to balance with or without the world (i.e., with or without humans), although it might not do so for a long time through evolutionary processes.

5 The writing of "anyone" having access is problematic with various issues of access, including, but not limited to, issues of digital divide and censorship.

6 A key connection with critical GC/E and PC/E is that ecopedagogical teaching widens persons' solidarity with all other humans and with Earth holistically, with a common goal of global and planetary inclusion of development and balance Misiaszek (2015, 2018b).

7 "Course" and "curricula" are in quotation marks to indicate that ecopedagogical spaces can be formal learning spaces, but also can be nonformal and informal education spaces in which both terms do not fit exactly accurately.

8 I have described the need for safe ecopedagogical spaces in the following quote:

> Ecopedagogues self-reflect on their strong environmental beliefs for students and teacher(s) to co-construct socio-environmental understandings and arguments together with everyone in the learning space. This requires ecopedagogues to engage in continuous self-reflectivity, to analyze and re-analyze how they teach and interact with their students. Safe learning environments free from violence (e.g. verbal, mental, physical, emotional) are essential for student(s)-student(s) and student(s)-teacher(s) conflicts on socio-environmental issues. (Misiaszek, 2020)

9 Reinvention of Freire is needed in problematizing his gendered terminology, such as "men," but also some of his later work did acknowledge his gendered terminology in his earlier writings including *Pedagogy of the Oppressed* Freire (2000).

10 Freire used Hegel's concept of Master-Slave dialect for determining hegemony, by reading and rereading the world (hermeneutics), as well by using Christian personalism, which focuses on love and mutuality (Morrow and Torres, 2002, p. 25). Hegel believed that the slave is the only one that can understand "true" freedom; but he also believed that pedagogy is needed for the slave to become conscious of this fact through revolutionary praxis. It is argued that such consciousness is essential for authentic citizenship that is based in democracy and freedom. Freire defined praxis as dialogical learning processes used to determine actions toward lessening oppressor-oppressed characteristic(s) of relationships (Morrow and Torres, 2002). Freire strongly agreed with Marx that subjectivism and psychologism is not true in the world, arguing that in the same way that "objective social reality exists not by chance, but as the product of human action, so it is not transformed by chance" (Freire, 2000, p. 51).

11 "Forcibly" is used because students are systematically not "given" the opportunity otherwise to express their own and others' realities.

4

Teaching for Ecopedagogical Praxis

Theories, Disciplines, and Positionalities

Education is itself a dimension of social practice. It seeks to know the reason for the practice and, through this knowledge as it deepens and develops, it also seeks new practice that is consistent with the overall plan for the society.

(Freire, 1978)

This chapter will delve into various aspects of ecopedagogical praxis that were discussed previously, with a focus on some key theoretical and disciplinary foundations, as well as methodological aspects of positionality and selections for theoretical and disciplinary frameworks. I will discuss some of the key issues of these topics with the realization that all of these aspects could be written about almost endlessly, as well as many other aspects of praxis that will also be missed. For example, the theories discussed will be incomplete in their description and analysis, and the list of theories is hardly all inclusive.

If critical pedagogies are grounded in Marxism then critical pedagogies are "vulgar" without praxis and, as critical pedagogies, ecopedagogies, are "vulgar" without praxis—as Moacir Gadotti (1996) has discussed:

> Inside Marxist thinking, this distinction seems to be clear. Certain distortions make Marxism, not a revolutionary instrument, but, rather, a conservative instrument. Among these distortions is the so-called academic Marxism, which is mechanistic and vulgar, which has no revolutionary sense, and is often of use just to show off learning.

To know how societies are oppressed by environmental violence and how our actions are making Earth increasingly unsustainable is not ecopedagogical work in itself, but rather to act accordingly through ecopedagogical praxis completes ecopedagogical work.

Dante wrote that the hottest, deepest circle of hell is reserved for those who know about injustices but do not act to end them. Scholars ((Pongratz, 2005) citing (Adorno, 1998)) have written that an increased recognition of the need for critical pedagogy emerged from two devastating events from the Second World War: the Holocaust and the invention of the nuclear weapon. Critical pedagogy raises two questions: First, how can we teach in such a way that students learn how to critically read and counteract human atrocities (i.e., crimes against humanity) when their leaders themselves are carrying them out? With respect to the Second World War this would be the Nazi, but there have been various acts of genocide since the end of the war. Second, how can we better understand each other so conflict will not lead us to the ultimate violent act of using an invention that can end the world and significantly destroy Earth overall[1]—the nuclear bomb? Environmental destruction, I would argue, is a second way in which we can cause both mass violence on the world and extreme destruction to Earth. This is despite knowing various ways that our actions are leading to this endpoint.

Unchecked, untethered environmental violence will ultimately lead toward the destruction of the world, with coinciding intensifying social violence as Earth becomes increasingly inhabitable for human. Without education that leads to solving environmental problems, the world will end in total environmental destruction, if it is not speeded along by the use of nuclear weapons. As previously stated, ecopedagogical practice and reading must be full of hope and dreaming of utopias; however, hope must be grounded on aspects of planetary sustainability and worldwide socio-environmental justice. It is within the realization that the world and the non-anthropocentric sphere cannot be separated in which we problem-pose the possibilities of solutions that our hopes conflict with ultimate balance for the rest of Earth.

As I have written in my earlier works through the research of expert ecopedagogues in Argentina (Buenos Aires, Córdoba), Brazil (São Paulo), and the United States (Appalachia region) (2011, Misiaszek, 2012), the ultimate in fatalism is when we, as humans, can no longer live on Earth, without a livable environment and resource to breathe, eat, and/or drink, with livelihood becoming the bare essentials at the base of Maslow's Hierarchy of Needs pyramid. Determining actions through deepened and widened theoretical reflection must also occur within the aspect of possible futures beyond our own lifetimes and the lifetimes of everyone "we" know currently. Hope must incorporate socio-environmental justice for future generations in which planetary sustainability is the baseline of any problem-posing discussion. If teaching is innately an act of hope for a better world for future generations with the aspect of continuing cultures, traditions, values, languages, and all else that make up communities (e.g., citizenships) and all that we cherish,[2] environmental pedagogies are essential for sustaining natural environments for such continuation,[3] as well as for bettering the world as part of Earth. Without teaching and research toward this goal, we will leave a hopeless world and Earth for future generations which, in turn, opposes the essence of teaching.

Some populations have less reason for such hope than others. Different societies and populations (e.g., based on gender, race, ethnicity, sexual orientation, (non)religious/spirituality, and nationality) are affected differently by environmentally violent acts and often these systematic socio-environmental inequalities are ignored in much of environmental teaching. As frequently viewed as too "politically sensitive" or "uncomfortable" topics they are not taught in neutralized schooling that constructs false apolitical classrooms. Such classrooms cannot be ecopedagogical because focusing on the struggles of those who are most oppressed is central to critical pedagogies (Gadotti, 1996). Ecopedagogues question how this lack of significant focus in environmental teaching on the differences of socio-environmental oppressions between populations helps to systematically sustain such gaps and often helps to widen them. Environmental pedagogical models, including ecopedagogical models, will be analyzed through theoretical framings (e.g., feminist theories, critical race theories, Indigenous theories, and colonializations (anti-, de-, neo-)).

Ecopedagogical usage of theories is for praxis toward the somewhat pragmatic goal of ending socio-environmental oppressions and planetary unsustainability (Misiaszek and Torres, 2019). If effective, seeing through theoretical lenses allows us to deepen and widen our understandings of others and their perspectives of socio-environmental violence to determine the "best" actions to take. This is with the realization that "dialectical thought, world and action are intimately interdependent" (Freire, 2000) but socially constructed (Scherr, 2005). It is only through such critical, dialectic deconstruction of world-Earth connections that actions can become humanistic, in that "action is human only when it is not merely an occupation but also a preoccupation, that is, when it is NOT dichotomized from reflection" (Freire, 2000). In other words, education is a cognitive tool for deepened and widened reflection for action, with the argument that "discovery cannot be purely intellectual but must involve action; nor can it be limited to mere activism, but must include serious reflection—only then will it be a praxis" (Freire, 2000). Without the goal of praxis, education is neither critical (Gadotti, 1996) nor ecopedagogical—but rather it resembles the goals of banking educations' reproductive teaching for oppressions and/or unsustainability to continue. Theories without this goal are virtually meaningless to employ because the world cannot transform without actions emergent from viewing the world-Earth through their lenses (i.e., ecopedagogical praxis).

Ecopedagogical theorizing helps to better understand the politics and sociohistories behind environmental violence and the positionalities of which populations benefit or suffer from environmental violence. Ecopedagogical selection and use of theories should go toward world-Earth de-distancing and de-distancing of "their" environmental oppressions from "our" environmental oppressions to collectively be viewed as "everyone's" environmental issues. With deepened reflection on everyone's environmental issues, a better understanding of the differences and commonalities within human populations and locations of Earth is a goal of ecopedagogical praxis, as well as having widening perspectives of everyone/being/thing makes up Earth. Key to this is to critically determine the Self's positionality within socio-environmental oppressions in relation to others' positionalities. The first topic to be discussed will be positionalities, which is an essential starting point of theorization for praxis and disciplinary selection.

4.1 Positionality problematizing

The key questions of positionality are as follows: What are the epistemological perspectives through which I see the world and those through which others view me? Paulo Freire's terminology of thematic universes could replace the wording of "epistemological universes" to have a similar meaning to construct generative themes. These questions, irrespective of the terminology, are not individually constructed but, rather, socially constructed, as are the epistemologies, too. Within more sociological lenses rather than psychological lenses, it is our Self within the world and, I would ecopedagogically argue, as part of Earth that has constructed our positionalities within the world and within the larger Earth. Positionalities are constructed by sociohistorical roles placed upon populations, which have no biological reasoning within truths; however, biological differences and commonalities can sometime provide false reasoning for positionalities. For example, I often remind my students that theories of feminism are based not upon biological differences but rather upon the roles that societies place upon those who have/identify with the gender of female.

The importance of deepened and widened reflection on positionality is to know the limits of the Self's positionality in understanding others' positionalities "from their own eyes." This section could delve into the complexities of feminist standpoint theories to critically deconstruct these limitations of "knowing" others' oppressions and the questioned viability of solutions, even through praxis, from others outside the population(s) suffering from the oppressions (Apple et al., 2009; Freire, 2000; Gadotti, 1996). For example, as a heterosexual, white, Western male, what are my limitations in understanding the oppression of others who identify as different in these aspects I have listed for myself? Queer theories, critical race theories, theories of whiteness, feminist theories, theories of colonialisms, and theories of globalizations (among others) can help me to better understand; however, there will always be various limitations in my understandings that will be (in)directly reflected in my teaching, learning, research, and critical readings. I pose the following question to my students every semester, although the exact details shift a bit over time:

> If I took my whole career to become an expert on female secondary school students in rural Afghanistan who are socio-economically poor within the local context—including reading everything I can find on the associated topics, utilizing various critical theories, and conducting research which includes me living within their communities for an extended amount of time—will I necessarily know "more" than a female secondary student who has lived in the community her whole life?

The answer is no, especially if it is a question of knowing the female student's own thematic universe (for this I would answer with an unquestionable "no"). There are ways in which I can utilize comparative techniques through looking at the proverbial fish in the fishbowl rather than being the fish, but I will never be able to fully understand as the female student does. With this stated, being the fish or being the observer of the fish has differing advantages and disadvantages in "knowing" the fishbowl but problematizing both positionalities is essential to better "know" the fishbowl.

Epistemological research must include the limitations of the research itself, especially in problematizing positionalities. Research for socio-environmental justice must give and/or amplify participants' voices on their perspectives, reasons, and possible solutions for the socio-environmental oppressions placed upon them. Problematizing positionality also includes the selection of research participants—that is, whose voices will be heard? Coinciding with the goals of critical pedagogies, if ecopedagogical research is to truly understand socio-environmental oppressions from those who suffer the most, research participants must be from these populations. Self-problematizing one's own positionality in the research is paramount because it is "I" who analyzes the data and limitations, as well as selects who will be research participants. All this problematizing of positionality must happen also within ecopedagogical learning spaces that are also spaces of research and through ecopedagogical teaching. Ecopedagogical teaching and reading must problematize the research given/read (including information and knowledges emerging from research) in relation to the participants, the methodologies employed, and the overall politics of the research, among many other critical questions. It is essential that there is always rigorous reflectivity on the limitations of positionalities within all research, teaching, and reading. Ecopedagogical work is erroneous when limitations and positionalities aren't sufficiently addressed.

Positionality within ecopedagogical work deconstructs the differing degrees of socio-environmental oppressions affecting different populations. Ecopedagogical theorizing helps all those in learning spaces deconstruct the differences between populations' social oppressions from environmental violence, in that there are commonalities between all beings part of Earth, but there are those who benefit from environmental violence and/or those who are less affected. In my last book (2018b) I problematized a quote from Ulrich Beck ((Beck, 1992) cited in (Hannigan, 2006)) who expressed that smog is democratic, but hunger is socioeconomic. I give the example of myself as a professor who, on days with severe pollution in Beijing, can choose to stay indoors and have the financial means to purchase air filters in my home and office. This is different from migrant workers who are outside for most of the day, suffering much more in the short and long term due to their socioeconomic positionalities. Critical theories help deconstruct the deepened and widened aspects of oppression, including the historical aspects that are often politically hidden.

Boaventura de Sousa Santos discussed the inequalities of oppressions throughout his work, with the following passage expressing the differences of positionality essential to epistemological analysis for deconstructing the limitations of knowing one's own positionality.

> Confrontational struggles may be of very different kinds. Each kind requires or privileges a certain way of knowing about the current society and the society for which one fights, about enemies or adversaries, as well as about the collectives engaged in the struggle themselves. (de Sousa Santos, 2018)

The privileges that people have are too frequently not fully self-realized; similarly, the Self cannot accurately position itself within the socio-environmental violence occurring. This is often witnessed by educators teaching critical topics that problematized students' privileges and is sometimes viewed by them as minimizing their "hard work"—for example, teaching critical race theories and theories of white supremacy in university to a mostly white classroom. This is especially true within ideologies of Western individualism in which "hard work" is seen as leading to achievement without inequalities as being relevant. This would be an example of how being an observer of the fish one can "know" things that the fish does not know about itself.

Problematizing one's own positionality, especially when one is not the oppressed, is essential to being part of the theoretical reflection for ecopedagogical praxis.[4] It is also essential that it is the historical as well as present structures of one's own identity that is deconstructed, beyond one's own actions. For example, the natural resources taken from the Global South, especially in unsustainable ways, is part of my Self as a Westerner, including additional origins of oppressions that arise from me being white and male. Ecopedagogical learning includes the deconstruction of the Self, as well as one's own histories of identities that lead toward socio-environmental oppressions and planetary unsustainability. Particularly with intensifying globalizations, global meaning making is necessary to understand the world and our positions within it, as Robert J. Tierney (2018b) defined it:

> Global meaning makers should, as Spivak (1988, 1990) suggests, be contemplative as they reconcile their complicity with their own privilege and adopt dispositions and approaches that are not presumptuous, colonizing, or recolonizing. The proposed model of meaning making requires self-interrogation of one's own enculturation in a fashion that involves continual scrutinizing of interests and activities and of positionality, perspectives, and biases.

Tierney (2018b) eloquently described the need for widened and deepened analysis of the politics, positionalities, and epistemologies for true critical global analysis. Such global meaning making is complex and conflictual; to "address the tug of war between homogeneity and heterogeneity, privilege and responsibility, global and local," with the oppressors making certain "not to position themselves as the savior or champion of others." Oppressors' might better recognize their oppressions upon others but complete class suicide is almost always impossible (á la Freire (2000), as discussed previously in Chapter 2, along with Memmi's (1991b) and Fanon's (1963, 1967) arguments on the limitations of oppressors' roles in liberating the oppressed) with the abandonment of one's own privileges as always limited.

Such arguments are more complicated within ecopedagogies in that the privileges of positionality the Self is expected to give up also make up the Self's identity, along with the complexities of planetary sustainability. Limitless complexity exists of the Self's positionalities within the planetary scope because in addition to oppressor-oppressed positioning within the world, there is the additional problematizing of the Self's positionalities with regard to planetary sustainability or unsustainability, which could or could not coincide with the ones of the world.

Reinventing Freire's (2000) call to counter the defining of "success" as the number of those "below" the Self, there is an ecopedagogical calling to counter the defining of "success" according to the consumption of the Self—of how much environmental violence is beneath the Self (playing off the metaphor of one's "environmental footprint"). Here I am talking about consumption of not only goods but also items including fuel (e.g., car, plane travel) and eating/cooking local food/ingredients.

Coinciding with (de)reconstructing success, positionalities as connected to ideologies of livelihood must also be problematized in ecopedagogical work. Livelihood of someone from the United States can be an example of such complexity, as it will be most often exponentially more planetarily unsustainable than individuals living in the Global South.[5] There are various complications to this argument such as diversity and wealth gaps within the two nations, but also pragmatically it would be impossible for everyone to live like people in the United States because it is inequalities that allow such livelihoods to exist—another essential aspect to problem-pose within ecopedagogical spaces. Returning to the praxis needed to define sustainability baselines discussed in Chapter 2, a key need for true planetary suitability includes the oppressors giving up their privileges from environmental violence seeing that the US baseline is unquestionably unsustainable.

But would this not be class suicide? I would argue that challenging the impossible-ness of class suicide is also necessary, with the recognition of the difficulties of "depreciating" one's own positionalities, as Freire had noted. Reduction of currently aspired-to livelihoods must be problematized within ecopedagogical spaces toward decreasing its associated environmental violence and unsustainability. My wording of "currently aspired to livelihood" does not mean lowering one's livelihood but rather reconstructing one's livelihood within justice and sustainability foundations so that, for example, it is not defined on neoliberal grounds. Such reconstructing of livelihood requires ecopedagogical praxis infused with problematizing what is valuable to one's life beyond success measured by those below and unsustainable consumption.

There are various aspects in which the positionalities of the Self and others will be problematized through various theories and pedagogies later in this chapter and overall in the book, such as through citizenships and education. For now, I want to state that within constructs of critical global and planetary citizenship, such global meaning making is at the heart of education to understand one's own role. This is achieved through endless self-reflection that also problematizes one's own thematic universe, sociohistorically influenced actions, and epistemologies that, in part, position the Self within oppressed-oppressor dynamics. Such self-reflection could result in praxis to deconstruct the historical divisions of "us" versus "them" within traditional citizenship models.

It is through critical self-reflectivity that we question what we know and how we know it as we position ourselves and try to position others in determining the politics with epistemologies of privilege in which "we" decide to commit specific acts of environmental violence or not. We commit numerous acts of violence without conscious thought, as I have within the past hour of waking up in Dar es Salaam as I use electricity for writing this sentence and used a hotel coffee machine that also disposes of a "convenience" plastic coffee pod, and as I will later take a taxi to a ferry traveling to Zanzibar that uses various fossil fuels and probably causes air and water pollution with its usage. In all these decisions, what reflection had I made and, most importantly, not made in these activities—as well as more than circulating Earth on a polluting plane to get here and other places for the past two months? What are the dynamics of positionality that "allow" me to have these acts of violence which, I would gather, is much more environmentally violent to almost all persons I can view from my window who call Dar their home? I do not know how Dar produces electricity but were there alternatives to the things I did since getting up that used no or less electricity? For example, if I justify to myself I could not have been "productive without coffee this morning," could I have brought my own instant coffee that would have not used the plastic container and that I would have known was produced within sustainable and social justice ways? With many, many more critical questions that can be asked, there is also the problematizing of how my privilege, as sociohistorically constructed, has gotten me to Dar to make these specific acts of environmental violence this morning?

The previous example gives an indication of the complexity of our own individual acts of environmental violence, in which actions within a relatively short amount of time can be extensively problematized. As my morning's actions will likely go unproblematized significantly, problematizing the larger systematic structures that lead to my environmentally violent action can lead to sustainability and less social injustices. This includes the questioning of my own positionality that allowed me to have the privilege of being able to commit these acts of environmental violence more easily as compared to most locals.

The need of critical global meaning is reflection, and also problematizing, while making sure not to position one's Self within a top-down process of false praxis, as Tierney (2018b) has written:

> Global meaning makers should be careful not to position themselves as the savior or champion of others . . . [rather] turn[ing] from self-righteousness to critical reflexivity, from imposition and imperialism to respect and restraint, as one develops cross-border understandings and challenges one's insularity, ignorance, and historic depreciation of others while interrogating complicity with one's own advancement via asserting forms of global jurisdictional mandates or misinterpreting, misrepresenting, or disregarding the rights, character, languages, and cultures of others.

"The tug of war between homogeneity and heterogeneity, privilege and responsibility, global and local" (Tierney, 2018b) needs ecopedagogical self-reflectivity within the world and, widening Tierney's quote, within Earth. Such self-reflectivity is never-ending in determining what we know and what we don't know, with self-reflection that this is difficult and that lapses of recognizing our privilege will occur. Ecopedagogical spaces must teach toward reading environmental violence of the world-Earth holistically, with the recognition of such reading for praxis from individual to large systematic levels. Such reflectivity is also on the limitations in such practices in continuing "toward," within the utopian education sense, absolute socio-environmental justice and sustainability as "the horizon" or "North Star." The limitations not only are within environmental classrooms but also include the limitations of teacher education in the theories in which form pedagogical practices are taught, for example, as de Sousa Santos (2018) has grounded oppressive epistemologies on histories of dominance. The endless complexities are discussed in Tierney's previous quote, with the necessary deepened richness and widened perspectives of critical global literacy and, I would also add, the "planetary meaning making" needed for ecopedagogical praxis. The radicalness of such praxis is the topic discussed in the next section, with the inclusion of de Sousa Santos' arguments of constructing transformative praxis within academics and education specifically.

4.2 Radical ecopedagogical praxis through diverse disciplinary, theoretical lenses

Critical theoretical and disciplinary (de)reconstruction for praxis within ecopedagogical spaces will be the focus of the rest of this chapter, with the realization that theoretical framing, through various theories, is contextual and limitlessly complex and, often, messy. As these are further discussed and exemplified throughout this book, I only have room to discuss a few key critical theories with rather brief descriptions that are limited to ecopedagogical aspects. By not enough "room" I mean to say there are limitless scholarship and contextual positionalities in which these theories can be described, as well as the numerous theoretical lenses that are needed to understand socio-environmental oppressions. Lastly, there is the aspect that was discussed earlier in which we have the utopian goal of understanding Earth and endless world-Earth connections through various epistemologies and "hard" sciences, social sciences, and the humanities, but it should be through the realization that full understanding will never be achieved (as a foundation of utopian thinking/teaching equated earlier as a horizon or the North Star).

Before going into the essence and essentialness of radical, revolutionary praxis within ecopedagogies, I want to reemphasize that praxis is being human and to deny it within education is dehumanizing, with domination only possible when praxis is denied. As well, theories emerge from human activities and "as praxis, it requires theory to illuminate it. . . . Human activity is theory and practice; it is reflection and action" (Freire, 2000). Praxis is hegemony's enemy because it questions the ideologies that sustain dominance. To best dominate someone is to do so not by strength of force but rather by manipulating them by hiding the domination itself, having the oppressed believe that what oppresses them empowers them (á la Marx). This is a foundation of banking education, which explains that empowering transformation is impossible through such pedagogies (Freire, 2000; Gadotti and Torres, 2009). I would argue with others, such as those I just cited, that banking education is a necessary tool for domination and sustainability as praxis is systematically denied.

Without radical disruption of the unchecked environmental violence meted out by the powerful upon the less powerful and the non-anthropocentric sphere, social injustices will intensify until "we" as humans will end up with either continuing social violence or environmental inhabitability of Earth. Although this might seem too pessimistic to some, I would question the level of violence that will erupt when the needs of those lying at the bottom of the pyramid in the *Maslow Hierarchy* become increasingly unavailable, leading to a fatalistic tipping point of the non-anthropocentric sphere that would cease to sustain the world without massive, or absolute, human extinction? If apolitical education was possible in which pedagogues teach the "facts" of the environmental crisis without political interference, I would question why we aren't all taking actions to end planetary unsustainability already. However, apolitical education is impossible (Freire, 2000, 1998c). With the term "radical" environmental pedagogies, in many ways it is not shifting the politics of teaching but rather teaching toward the truths of how human actions are within the nonsubjective laws of Nature without political and ideological "interference."

Radical ecopedagogical praxis is achieved through valuing all of the world's populations holistically toward planetary sustainability. Oddly enough, ecopedagogical praxis should not be termed as "radical" but named as such because it is countering politics and ideologies to sustain hegemony, through such ideologies as neoliberalism, from inequalities and planetary unsustainability (this could relate back to Hegel's Master-Slave dialect as discussed previously). The subjectivity of the world's epistemologies of the non-anthropocentric sphere within sociohistorical oppressions is a key point of tension in which planetary unsustainability emerges with always coinciding social violence. Social oppressions emerge from both unequal environmental violence upon one population(s) as compared to another and environmental unsustainability.

Ecopedagogically questioning the ideologically unquestionable brings to light how environmental violence is justified as violent and unsustainable, while hiding the true politics. The ideology of neoliberalism, which Postma (2006) has argued as only caring about one's own private sphere of power and related consumption, frequently becomes politically acceptable to those who are worst affected. An obvious current example of this is Trump getting elected and his strongest supporters are largely a population with demographics that will be most hurt by his neoliberal, antienvironmental policies. Radical pedagogies and actions have increasingly been on the defense, even from populations that are most oppressed.

> Most morally repugnant forms of social inequality and social discrimination are becoming politically acceptable. The social and political forces that used to challenge this state of affairs in the name of possible social and political alternatives seem to be losing steam and, in general, appear to be everywhere on the defensive. (de Sousa Santos, 2018)

A key aspect of ecopedagogies, as a Freirean pedagogy, is being radical, with the unlearning that "radical teaching" or "teaching for radical action" is negative. As de Sousa Santos discussed the "losing steam" of transformative possibilities, a key question for education is how it would be utopic, as described previously, in that oppressive normative systems/structures are not fatalistically infinite without possible change. Those who benefit from socio-environmental violence rely on "acceptance" without protest or alternative thinking.

Radicalness can be seen as countering what is invoked in the term of "common sense." Common sense is plural in meaning, with Carlos Alberto Torres (2011) giving it two key characteristics in the passage below in relation to "truth."

> One is that common sense becomes "naturalised" in the lives of people as something normal that we do or ought to do just to get by and/or prosper. Common sense is a generalized truth about something; it is the normal way to do things, the normal way of becoming human beings. Secondly, the idea of common sense is based in culturally shared understandings or values. That means that what is common sense for some people may not be common sense to others. Therefore, we ought to consider the notion of common sense as variable across cultures . . . let us clarify a potential conundrum: how we differentiate common sense from Foucault's concept of "regimes of truth." In my view the two concepts are different. For Foucault, a regime of truth is a historically specific strategy that produces discourses functioning as (or claiming to be) truthful in specific times and places.

Torres writes on problematizing common sense especially within intensifying globalizations from above, as the above quote is from an article on public universities and globalizations that normalize neoliberal ideologies (Torres, 2011). One of his main arguments is that the depth of neoliberalism as unquestionable common sense is increasingly de-contextualized in educational systems, especially at the tertiary level.

Ecopedagogical teaching and literacy need to challenge oppressive common sense that serves as ideologies for continued unjust power and inseparable connections to planetary unsustainability. Utilizing the work of Haluza-Delay (2003); Tierney (2018b) wrote on the need for praxis to emerge through knowledges of oppressions to determine a solution.

> Haluza-Delay (2003) asserts: "Knowledge uncovers the oppressive structures and confronts power. However, it is not the 'knowledge' alone that does this, but the process by which the knowledge is taken up and used in the community, altering 'common-sense.'" In accordance with activism, global meaning making involves "thinking, talking about, researching and theorizing

> about what is going on, what they are going to do next and how to analyse the situations they face, whether in relation to attending a demonstration, a meeting, a confrontation with institutional forces or planning the next action or campaign." (Kinsman, 2006)

This has been stated previously in the book as a critical and coinciding ecopedagogical goal, but the pragmatism with which Tierney utilizes this within his arguments of deepening and widening global meaning makes it essential here.

To reemphasize the previous argument on "radical" labeling, teaching that counters such socio-environmental violence is inherently pedagogically radical because ecopedagogical teaching questions and often counters many ideologies that are unjust and unsustainable, but are dominant within educational structures and society overall. For example, ecopedagogical work counters neoliberalism in the West which is spreading through globalizations from above, because it is a root cause of socio-environmental injustice and unsustainability. Thus, teaching for praxis through problematizing whether there are alternatives to neoliberalism is inherently "radical." Not only such "radical" praxis but also the roles of ecopedagogues themselves should be taught. As Apple and Au (2009) have argued that critical pedagogues must be aligned with radical movements and activists whose goals are for transformation within social justice, ecopedagogues' roles must align similarly.

Ecopedagogical teaching for "sustainable development" (i.e., ES*d* teaching) centers on the questioning of "development" itself—earning its "radical" label—unlike ES*D* models that work toward static, non-contextual "development" goals, often for "them" rather than for "us," "everyone," and planetary sustainability. ES*D* is reproductive education of *D*evelopment rather than *d*evelopment for all the world, because *D*evelopment is not deconstructed, and *d*evelopment is not the product of reconstruction. Radical ecopedagogical praxis toward *d*evelopment is the goal, with the recognition again that knowledges learned within ecopedagogical spaces are useless against *D*evelopment without such praxis.

I end this section, before summarizing the following subsections, by questioning if ecopedagogical radicalness can be within formal schooling, especially up to the secondary level. This question emerged as a central theme in my ecopedagogical research in Argentina, Brazil, and Appalachia (the United States) (Misiaszek, 2012, 2011). This question could be asked about all critical, transformative pedagogies with answers that might lead to the follow-up question of "what degree of radicalness?" The essence in answering the overall question lies in determining whether a teaching and learning that continuously questions the very system it is part of can survive. Radical systems of pedagogies, including environmental pedagogical systems, are necessary in which radicalness is essential from "within the system" that allows for ecopedagogical learning spaces that question its structures. Do, and would, most schooling systems allow for this?

The next subsections discuss how ecopedagogical work is transdisciplinary and trans-theoretical, and the next chapter will argue the trans-epistemological essence of ecopedagogical work. All three of these are impossible to separate from each other so there will be quite a bit of overlapping, but each section will focus on more specifics of each independently.

4.2.1 Transdisciplinary constructions

> Transdisciplinarity is "more than a new discipline or super-discipline, it is actually a different manner of seeing the world, more systemic and more holistic" [(Max-Neef, 2005)], introducing a kind of quantum logic as a substitute for linear logic [(2005)], and breaking with the assumption of a single reality (2005). Max-Neef (2005) suggests we view discipline and transdiscipline as complementary, as transdisciplinarity rests on a coordination of empirical, pragmatic, and normative research and is constituted by explicit values, ethics and philosophical positions that extend beyond disciplinarity. (Shultz and Jorgenson, 2009)

The world and Earth cannot be deeply and widely understood through a single discipline or a set of disciplines within a singular epistemological framing, but rather understandings must be through multiple disciplines of learning. Lynette Shultz and Shelane Jorgenson (2009) explained in the quote, through Manfred Max-Neef's (2005) scholarship, what is transdisciplinary and its need in the teaching of critical global citizenship in higher education, but such transdisciplinary-ness is essential for ecopedagogical work in all types and levels of education. The discussion on the need for transdisciplinarity within ecopedagogical work can be endless; however, this section will briefly discuss the following three aspects within ecopedagogical work: whole-curricula approaches, non-technocratic approaches, and sociohistorical and epistemological problematizing of disciplines selected. In addition, there have already been and there will continue to be discussions on ecopedagogical transdisciplinarity before and after this section.

First, as I have argued elsewhere through research (2018b, 2015), ecopedagogies must be through all the curricula and must also be emphasized in their courses within the curriculum. Ecopedagogical teaching must be focused within courses throughout the curriculum but also must be meaningfully and thickly threaded throughout all courses, including the social sciences, the "hard" sciences, and the humanities. Some scholars call this whole-curricula approach when discussing larger topics that are essential for world well-being such as GCE, Planetary Citizenship Education, and peace education, as a few examples. Ecopedagogical curricula must be transdisciplinarily constructed with all those in the learning space with the goal of diverse understandings that are often conflictive with one another and always endlessly complex. There is always a limitation in scope of what ecopedagogical learning spaces can incorporate; however, there are much more severe limits if only taught with a single course or a set of courses rather than throughout students' curricula.

As Freire stated in his writing, learning and praxis is hindered by teaching that is segmented, disconnecting and distancing lessons from societies outside the classroom and other learning spaces; the mystifying Nature of such segmented teaching does not lead to *d*evelopment.

> Completely coherent with this vision is the concern for replacing the concept and practice that sees each part of the system as a separate entity. When each segment of the system is isolated from the others, the learner's own development is forgotten and each stage becomes an alienating and alienated moment, merely a preparation for the next stage. (Freire, 1978)

As I argued in my previous research (2018b, 2015, 2012, 2011), environmental learning often takes place in artificial disciplinary vacuums, or silos, away from other disciplines. However, it was also found in my research that is not by mistake but also politically constructed systematically for suppressing authentic dialogue and emergent radical ecopedagogical praxis to mystify socio-environmental injustices and unsustainability. This coincides with Freire's (1978) wording of self-development as forgotten and alienating from curricula segmentation. The disconnections of such mystified curricula are to lessen/remove possibilities of praxis, especially for development. Without self-reflection or resulting praxis, teaching becomes technically stepwise as "merely a preparation for the next stage" rather than deepened and widened learning through *and* between all stages.

My second argument is that technocratic environmental pedagogies do not allow for meaningful connections to the world-Earth and, thus, that teaching for praxis toward socio-environmental justice, *d*evelopment, planetary sustainability becomes nil. Let's take for example the "hard" sciences with its problematic "hard" labeling. In technocratic environmental pedagogies hard sciences are separated from all other disciplines as though hard sciences' knowledges, practices, and research are within a vacuum without being influenced by politics, biases, subjectivity, epistemological commonalities/differences, and other disciplines' knowledges, among other influencing factors. Although the laws of Nature are objective, this does not mean, nor does it happen as many scholars have argued, that sciences, disciplines, research, and "official" knowledges are apolitical or objective in the absolute sense.

For example, Freire discussed how global economics of being without labor is beyond those of the disciplines of economics and business, but rather includes those of the hard sciences and accompanying technologies emergent from it.

> As I have said before, worldwide unemployment is not a fatalistic inevitability. It is the result of the economic globalization and the scientific and technological advances that lack a form of ethics that serves the interests of all human beings and not just the unfettered greed of the power minority who control the world today. The application of technological advances, which requires the sacrifice of thousands of people, is one more example of how we can be transgressors of a universal human ethic in the name of the market, of pure profit. (Freire, 1998b)

Although not stated directly, it can be inferred that global economics are affected by sciences with the need of the social science and humanities to problematize ethical questions as they are perverted by being guided by monetary profit within neoliberal ideologies. Without utilizing a transdisciplinary approach to understanding local unemployment, understandings of economic/labor oppressions would not be sufficiently understood for effective solutions, with only superficial solutions emerging without significant or sustainable relief. Within ecopedagogical spaces, transdisciplinary problem-posing of environmental violence and unsustainability would include such aspects as how sciences have increased or decreased the intensity of, the spread of, and the distribution of, environmental violence. Ecopedagogical work includes the current aspects of science, as well as the histories of sciences and resulting technologies, in which their uses are contested terrains of both oppression and empowerment, depending on the context. These arguments against technographic environmental pedagogies also emphasize the need for the holistic approach discussed earlier in this section.

Another example is the internet which allows for global communications and the distribution of knowledges, which can bring either diverse knowledges and voices for understanding environmental issues or opposingly narrow understandings by feeding into individuals' fears and prejudices through manipulated information and search results. These are two of numerous ways the internet can narrow and falsify environmental understandings, as both are intensified within post-truthism (see Chapter 7).

The important aspect argued here is that sciences and technologies (which emerge from sciences) must be taught through transdisciplinary lenses. False disciplinary separation dehumanizes sciences and leads to practices and inventions (e.g., technologies) from sciences not having humanizing processes, reasonings, and results/products. As Freire argues it is not teaching against hard sciences or technologies, but humanizing them, or I would add planetarizing them also within ecopedagogical tenets.

> [It] does not mean that we are to adopt a false humanist posture of denying the value of technology and science. On the contrary, it's a posture of balance that neither deifies nor demonizes technology. A posture that is from those who consider technology from a critically curious standpoint. (Freire, 1998b)

Freire objects to not problematizing sciences through multiple disciplines to determine points of empowerment and oppression, and humanizing and dehumanizing. Ecopedagogical spaces focus specifically upon how such manipulation of, isolation of, and apoliticization of disciplines counter the deepening and widening of environmental understandings, inside and outside learning spaces. This learning through critical deconstruction needs transdisciplinary teaching and reading for both "a posture of balance" and from "a critically curious standpoint" of humanization of the world-Earth holistically.

Third and last, ecopedagogical work must also deconstruct the "histories of the disciplines," as emphasized by various critical scholars (Apple, 2014, 2004; de Sousa Santos, 2018, 2012b, 2008), including the historical constructions of their curricula and "official knowledges." Such sociohistorical deconstruction of disciplines was found as essential in ecopedagogy in my previous research work. Utilizing the work of de Sousa Santo and Raewyn Connell, this will be further deconstructed in the next section on the trans-theoretical needs of ecopedagogies, as well as trans-epistemological aspects in the next chapter, which problematizes theories emergent from sociohistorical oppressive epistemologies that only reproduce the oppressions. However, scholars often use these same theories to end oppressions, as de Sousa Santos and Connell argue that it is impossible to counter the very epistemological foundations of the theories and the disciplines the theories emerged from.

The essence of this problematizing of (trans-)disciplinary work is essential in ecopedagogical work, including the theories, on which these disciplines are historically grounded, that de Sousa Santos (2018) discussed as the sociologies of absences, as that which is not part of the canon of sociology. De Sousa Santos (2018) calls for the radical reinvention of disciplines such that historically oppressive elements are removed from the disciplines allowing the "emergences" of disciplines absent of socio-environmental oppressive foundations to truly lead to ending oppressions. Within sociology, he termed such radical reinventions as sociologies of emergences (2018).[6] Such critical deconstruction and reconstruction of disciplines are impossible within the discipline alone, but rather need transdisciplinary deconstruction within and between disciplines (Misiaszek, 2019).

4.2.2 Trans-theoretical framings

> This critical reflection on your praxis is absolutely indispensable. It should never be confused with meaningless alienating and alienated talk. While it is the source of knowledge, praxis is not, however, a theory in itself. It is only when we give ourselves constantly to critical reflection on it that praxis makes possible the development of theory, which, in its turn, illumines new practice. (Freire, 1978)

Theories must not be stagnant but fluid as contextually meaningful tools to better understand "what is going on," with critical theorizing as focusing on how to better understand oppressions for transformative praxis. As will be explained in the next paragraph, critical reflection uses theories to "illumin[ate] new practices" rather than to mystify, with praxis as utilizing theories as tools for deepened and widened reflection. Theories need to be contextually meaningful for those in the ecopedagogical spaces, while the theories' foundations remain intact. Theories used in ecopedagogies should be contextually reinvented, but not apoliticized or de-radicalized, and ecopedagogies should not ignore a theory's essence—if this is the case the theory should not be used. Freirean work is for radical transformation to counter oppressive politics through teaching to understand the politics of social injustices, and ecopedagogical work is for radical transformational praxis through focusing on the politics of socio-environmental violence and widening understanding of politics that lead to planetary unsustainability (Gutiérrez and Prado, 1989; Kahn, 2010; Gadotti, 2008a, 2008c; Misiaszek, 2018b, 2011). Usage of theories within ecopedagogical work should be for progressing toward these goals. In other words, it is ultimately for understanding without dilution and theories should ultimately be pragmatic and practical because they would be useless otherwise, but their pragmatic-ness is often hidden. When politically hidden it equates to the denial as Freire had discussed in the previous quote.

In my classes, my utterances of the word "theory/ies" often terrify students, while undergraduate students show their fear on their faces outwardly, graduate students are able to hide it a bit more. As a student at these levels, it would be the same for me. I often also hear in and outside my classes that they don't like or are "sick of" theories, because they like "practice." I try to begin my classes and accompanying readings by asking: What is a theory or what is a framework of theories? In short what are they for and how have they come about. I find that theories are often mystified as only used by academics who force their students to read them, while using complex terms to satisfy their self-importance (i.e., ego). Theories' descriptions and definitions often include numerous words that end with "ism," which leads to further mystification rather than better understandings. I am not arguing that theories should be superficially simplified or narrowed; on the contrary, I argue that theories should be meaningful and contextual for those using them to understand the complexities and messiness of the world-Earth.

In a previous book (2018b), I discuss how I, along with PowerPoint slides with various Nikon lenses (from fisheye to ultra-zoom lenses), tried to illustrate to students that theoretical lenses help better understand the world and Earth. However, do theories help to understand "realities" and others' perspectives of their "realities" or rather mystify them? Theories must be toward demystifying realities by, in part, actually emerging from realities, for ecopedagogical praxis to happen.

If education does not proceed toward this end, with accompanying theories, the theories are useless and education is dehumanizing and deplanetarizing. Opposite to this goal, banking-education models are purposefully structured toward such dehumanization, as Freire (2000) argued: " The dominator has no choice but to deny true praxis to the people, deny them the right to say their own word and think their own thoughts." The mystifying of theories is a form of denying praxis to read the world and Earth, and thus helps to reproduce oppressions and unsustainability.

4.2.2.1 Theoretical selections

A good question is the following: What theories are necessary for such radical ecopedagogical praxis? Without going into the specific reasoning of theories that I have selected to utilize in my own ecopedagogical work, on which I have written extensively through many other publications (2018b, 2011)[7], I will discuss some ecopedagogical tenets for theories selection. Theoretical selection calls for the epistemological interrogation of theories to ask if lens disrupts and de-distances unsustainable, socio-environmental violence. Interrogation also includes problematizing if the theories (and epistemologies) are actually "doing" the opposite—helping to sustain/intensify socio-environmental oppressions and unsustainability. If they are, another question is if the theories could and should be reinvented by, in part, determining if the theories' foundations are useful for radical ecopedagogical praxis.

Radical ecopedagogical praxis cannot occur within ideologies of success quantified by how many are "below" one's Self, and thus, liberation of the oppressed is becoming the oppressor (problematized by Freire, as previously discussed). Within the three ideologies of sociohistorical analysis that de Sousa Santos utilizes in his work (i.e., coloniality, patriarchy, and capitalism), "success" can be connected to other aspects previously deconstructed, such as d/Development, the Self's livelihood with/out others' livelihoods, and solidarity/othering of citizenships. It is essential to know the theories and reinventions of theories that are not grounded in such oppressive foundations to problematize for theories selection for ecopedagogical work. Or, in other words, what theoretical frameworks are necessary to counter the oppressive and unsustainable constructs of, for example, the aspects given in this paragraph? This requires rigorous sociohistorical deconstruction of theories to determine ecopedagogically appropriate usage. Without giving specific arguments for the appropriate usage of theories, I will state that thorough reading and rereading of these aspects, among many others, must be done. Ecopedagogical theorizing should center diversity in understanding diversity and counter theories, and aspects of theories, that de-diversifying the concepts of Earth and world-Earth connections, including neoliberalism, Development, and othering citizenship constructs. Also, theoretical selections must also identify and deconstruct the socio-environmental fatalism emergent from these aspects, especially for ecopedagogical praxis toward ending their reproduction.

Within ecopedagogical spaces, the theories utilized in teaching and reading the world-Earth are selected according to what is needed for learning about socio-environmental connections to violence and unsustainability. As previously stated, there is no single or single set of theory(ies) but rather theories are democratically and bottom-up selected with the goal of better reflecting on the best possible solutions to end unsustainable environmental violence, especially to those who suffer the most. Within Freirean terms, which were discussed previously, what theories are necessary to teach, and read through ecopedagogical literacies, to overcome limit situations determined by the current and utopic socio-environmental Earth we envision? Ecopedagogical spaces have the additional aspect of theorizing beyond the anthropocentric sphere to understand the non-anthropocentric sphere in connections within and outside the anthropocentric sphere.

Throughout this book already I have emphasized and reemphasized that ecopedagogical teaching is deepening and widening understandings from those who suffer the most from socio-environmental violence and unsustainability. Theoretical framings and selection must ground this goal. The question of who suffers the most is a question that needs to be theorized in itself: Which theoretical lenses do we "look through" to understand others' socio-environmental suffering?

In the same respect of using theories to better understand oppressions from those who suffer the most, the sociohistorical foundations of theories must be problematized in theoretical selection. Problematizing if theories grounded on socio-environmental oppression can provide praxis for ending such oppression is essential (Takayama et al., 2017; Connell, 2013, 2007; de Sousa Santos, 2018, 2007b). But also, there are the moral and ethical aspects: whether theoretical building from/upon oppressions or other theorizing should be utilized to end it. This includes theories termed as "critical" that are often utilized to counter oppressions. De Sousa Santos argues that "critical theories" frequently serve to solidify a single epistemology that is sometimes grounded in accepting colonialism, capitalism, and patriarchy.

> Critical thinking, seems to have been much more successful—so much so that the gradual narrowing down of the alternatives laid out by progressive critical thinking has reached such an extreme in our time that it becomes possible to say what in the last two hundred years was considered too patently wrong to be said: there is no alternative. (de Sousa Santos, 2018)

De Sousa Santos (2018) in this quote and throughout his writing, is not referencing critical theorizing and pedagogy of bottom-up models such as Freire's work, which he later discussed extensively to construct his self-termed "Postabyssal Pedagogies," but rather critical theoretical foundations of "Western-centric critical thinking," especially without reinvention.[8] He continues his argument that use of such Western critical theories, with the foundation of transformation within capitalism, colonialism, and patriarchy, has "g[iven] rise to an immense historical frustration made of perverse effects, dreams sliding into nightmares, hopes ending up in deeper fear, and revolutions betrayed; civilizational gains deemed" (de Sousa Santos, 2018).

Ecopedagogical theories selection should include theories outside of those grounded in epistemologies of the North because their overall ways of knowing relies on separating the world from the rest of Earth. Theories that epistemologically counter and/or widen beyond Western reductionism and objectification allows for a holistic shift in interrelatedness within the world, the world within Earth, and the spirituality of and beyond Earth (Tierney, 2018b). Using explanations of some key Indigenous knowledges' aspects described by Tierney (2018b), he explained them as "experience, observation, history, language, stories, dreams, Nature, and animals; the Nature of the state of knowledge—its animations, permanency, or changeability; its state of flux; and its relationship to the past, present, and future" (Tierney, 2018b). Planetary understandings from such concepts will not emerge from epistemologies of the North, as de Sousa Santos defines them.

In ecopedagogical research, there is the selections and framing of multiple theories that also emerge from the research field directly—from the voices of the participants. Theoretical selectivity is one of the many reasons for the need for researchers from the places that are being researched, that is, researchers from the Global South, rather than the current predominance of researchers from the Global North in researching the Global South. This does not mean that research from those outside the local contexts is always negative, but it definitely doesn't mean that just research from the dominant North, by Northern researchers and through Northern epistemologies, theories, and methodologies is "good." What can be stated is that research that disregards contexts and local methodologies is top-down—and hence neither critical nor ecopedagogical. Coinciding with these arguments, rigorous self-reflectivity and positioning of the Self's epistemological framing is necessary within ecopedagogical theorical selection to determine the needs of utilizing theories outside of one's own epistemological groundings.

Lastly, I return once again to post-truthism with the argument that ecopedagogical theoretical selection and reinvention must counter post-truthism—this is needed more than ever before. Without specifically discussing post-truthism, de Sousa Santos (2018) gives the following poignant argument of epistemological politics, in which questioning the possibilities of thinking different is famed as false: "Political becomes epistemological when any political alternative to the current state of affairs is credibly framed in the same way as fancy against fact or as falsehood against truth" (2018). Although this could be taken that post-truthism should be considered as just "alternative ways of thinking" or "alternative epistemologies," false truths emergent from post-truthism are the politics of lying to indoctrinate opinions as truths, rather than epistemological diversity toward better understanding through emergent truths.

4.2.2.2 De-coloniality

Ecopedagogically reading the world today must be through the histories of colonialisms to understand world-Earth injustices and unsustainability, especially with intensifying globalizations from above, which are appropriately linked to the term "neocolonialism." Ecopedagogical work calls for the radical disruption of coloniality and for a transformation absent of it. Stein has argued the following needs for decolonizing teaching which calls for disrupting epistemologies that continue colonial legacies in reproducing social oppressions and fatalistic futures.

> The first move is to interrupt and denaturalize the epistemological and ontological frames of modern existence that are produced through violent and unsustainable relationships and processes—and to ask about the role of education in reproducing and/ or interrupting these frames. This means going beyond simply rearranging the content within existing frames, and instead stepping back to ask about the conditions and productive effects of the frame itself. The second move is one of grappling with/at the limits of the kinds of relations and futures that are possible within existing frames, so that we might start to disinvest from attachments to their harmful promises, and to learn the lessons from our repeated mistakes. Finally, the third move is one of attending to the need to reimagine and recreate what is possible—and to ask about the role of education in facilitating these other possibilities. (Stein, 2017)

Radical disruption of coloniality in the world's ways of knowing that reproduce environmental violence and normalize the resulting social violence, dominance of the rest of Nature, and planetary unsustainability is foundational to ecopedagogical decoloniality.

Decoloniality theorizing (or anti-coloniality or postcoloniality, depending on how the three are defined) deconstructs how histories of colonialism continue both directly and indirectly today and into the future for praxis toward removing all residues of colonialization. An initial obstacle that decoloniality theories help learners to understand is that the argument that coloniality is no longer relevant is a myth. The following quote by Grosfoguel (2008) expressed the power of this myth and the reasoning for using the term "coloniality" rather than "colonialism" to help counter this myth.

> One of the most powerful myths of the 20th century was the notion that the elimination of colonial administrations amounted to the decolonization of the world . . . le[ading] to the myth of a "postcolonial" world. . . . Herein lies the relevance of the distinction between "colonialism" and "coloniality." Coloniality allows us to understand the continuity of colonial forms of domination after the end of colonial administrations, produced by colonial cultures and structures in the modern/colonial capitalist world-system. "Coloniality of power" refers to a crucial structuring process in the modern/colonial world-system that articulates peripheral locations in the international division of labour with the global racial/ethnic hierarchy and Third World migrants' inscription in the racial/ethnic hierarchy of metropolitan global cities.

Ecopedagogical use of decoloniality theories is to deconstruct the "coloniality of power" that leads to intensified socio-environmental violence for the Global South, including and beyond aspects of labor that Grosfoguel exemplified in the quote above. When students and colleagues sometimes question if colonialism's residue still significantly affect current education and environmental issues, I ask them to overlay a map of the world showing past colonialism with a current map to see how same patterns of global power inequalities, including socioeconomic ones, and differing degrees of environmental devastation exist in the two maps. Another question I pose to them is this: Why are many of the most natural-resource rich nation-states some of the least globally powerful in various ways and most environmentally devastated—what are the histories of colonialism for those "nations-states"?[9]

For example, how do histories of colonialism continue to oppress the vast majority of South Africa with non-whites making up the vast majority of the population but accounting to an insignificant percentage of the nation's economic wealth? Which part of South Africa's population benefited from the once most valuable appraised lands in the world due to diamonds lying underneath (Meredith, 2007), and from the vast gold mines?[10] Of course, this question is best asked as who outside of South Africa's population benefited from such mining. Beyond economic and livelihood questions, other questions that are relevant today pertain to how structures of South Africa's societies, very much including education, have lasting negative effects, including unsustainable socio-environmental violence. The university students–led movement of Rhodes Must Fall is an excellent example of decolonizing praxis to remove all coloniality from Cecil Rhodes' legacy with the very pragmatic aspect of removing his statues and other imagery from campuses, as well as throughout South Africa and England.

Within ecopedagogical learning spaces and through ecopedagogical literacies, there is an essential need to theoretically interrogate epistemologies of coloniality on how injustices and unsustainability are sustained due to their histories of ideologies. In line with Albert Memmi's quote of us "hav[ing] no idea what the colonized would have been without colonization, but we certainly see what has happened as a result of it" (1991b), a key question would be: What South Africa would have "been" if it had not possessed a rich source of diamonds and gold? However, a better question might be what South Africa would have "been" without colonialization due to their global location, which largely lead to colonizers "discovering" the rich resources of diamonds and gold? This second question is from colonial history that before the "discovery" of diamond and gold, the global location of South Africa at the tip of the continent initiated colonialization especially during the time the UK was worried about impending conflict at the Suez Canal impeding on ocean navigation, which, in turn, would hinder the transportation of resources taken from other colonies (Meredith, 2007).

This thickened sociohistorical analysis exemplifies the layers of depth essential to deepen and widen the influences of coloniality as both, interconnectedly, causes and effects of environmental violence. It is impossible to understand all the messiness of coloniality; ecopedagogical teaching and reaching must be, as Tierney wrote on global meaning making, "alert to the colonizing constructions of the world (e.g., hegemonic, ethnocentric, ahistorical, depolitic, paternalistic, simplistic) and our complicities with their systemic continuation" ((Tierney, 2018b) citing (Andreotti and Souza, 2012)). The interrogation of the world as apart from Earth is essential in decolonizing ecopedagogy in that colonization has largely placed Nature to be profiteered from through economics of coloniality but also within dynamics of power in various ways, many of them purposely veiled.

Within the field of education there are questions of how the world is taught, as well as how do we research education—through what theories, what disciplines, and what epistemologies? Decoloniality is essential in answering these questions through and between various theories, such as the following: Edward Said's (1979) theories of Orientalism in that all the world is viewed through Western lenses; theories introduced as internalized colonialism by Fanon (1967, 1963) and Freire using Fanon and others; and the epistemological work of de Sousa Santos (2018, 2014, 2007b) and Raewyn Connell (2014, 2013, 2007) of deconstructing and delegitimizing epistemologies of the North through epistemologies of the South. Decolonial problematizing on how education, especially higher education, allows for the theorizing of the world and Earth—with particular focus on what theories or uses of theories are absent (Tierney, 2018b) ((Takayama, 2009) cited in (Tierney, 2018b))—is necessary. Such problematizing is essential because "we," as academics, often "represent forms of enculturation that embed Southern or Eastern studies into Western frames" (Tierney, 2018b). This also includes the foci and methodologies of research "we" do.

What epistemologies have we dismissed within academia utilizing "our" theories, and the world overall, because they do not "fit" into dominant Western knowledges and ideologies, which have led to socio-environmental injustices and planetary unsustainability? Ecopedagogical work that problematizes the epistemologies that "condone" and often "encourage" violence, questions of decoloniality of development (*d*/*D*evelopment), sustainable development (S*d*/S*D*), and its education (i.e., ES*d*/ES*D*) is essential. Various scholars link back constructs of degrowth as providing alternatives to *D*evelopment that "decolonize[s] the imaginary (of development) from growth" (Singh, 2019; D'Alisa et al., 2015; Latouche, 2004).

If most "development" has the endpoint of becoming more "modern" through processes of modernization (Arnove, 2007; Cudworth, 2003), how do we define modernity and what epistemological foundations do we understand it through? As stated throughout this book, ecopedagogical praxis is toward multiple modernities, with the plurality of modernit*ies* to signify multiple framings and contextualized goals, in opposition to *D*evelopment's single modernity endpoint (Misiaszek, forthcoming, 2020). Modernities must be problematized in the ways in which Western-Eurocentric modernity (i.e., the Global North) have entrenched "movement[s] of subalternisation and normalization is generally referred to as the 'epistemic violence' of colonialism" (de Oliveira Andreotti, 2011b). De Sousa Santos (2018) argues that many decoloniality scholars have focused on how colonialization represents broken European promises of modernization, but also what is necessary is "cross-cultural dialogue without the baggage of imperialism . . . [i.e.,] the dialogical side of decolonization" (2018) in how we frame the colonial ways of knowing "development" with goals of "modernization." Slightly rewording de Sousa Santos' argument in the form of a question: How can we counter *D*evelopment and modernizations that emerge from epistemologies of the North grounded in colonialism, as well as patriarchy and capitalism? Ecopedagogical dialogue and reading through decoloniality theories are essential for deepening and widened and widening sociohistorical deconstruction of *D*evelopment, and modernity is essential to determine *D*evelopment, "without the baggage of imperialism," with theorizing grounded in epistemologies of the South to further widen lenses to the planetary sphere. It is important to state that de Sousa Santos (2018) not only writes such arguments for noncritical environmental pedagogies but also argues against the historical "undercurrent of Western modernity" traces of "philosophy of Aldo Leopold and the deep ecology of Arne Naess" (Naess, 1990; Leopold and Brooks, 1933).

Decolonization of theorizing is essential, which must be beyond "epistemological diversity" that only returns to Western frameworks. If epistemological diversity is being sought, then current practices must serve different ends, countering global trends of practices representing "forms of enculturation that embed Southern or Eastern studies into Western frames" (Tierney, 2018b). The following quote by Keita Takayama given by Tierney (2018b) discusses the responsibilities of scholars from the Global North:

> Given that the existing unequal structure automatically warrants Western scholars the right to speak "on behalf of the world," they have ethical responsibility to bring in sophisticated theoretical work from the margin that should immensely contribute to the discussion . . . in the centre. . . . Democratic space must be generated . . . where non-Western scholars and activists can participate in theoretical knowledge production on an equal footing with Euro-American counterparts. (Takayama, 2009)

I very much contend that ecopedagogies from the Global North must adhere to the arguments above.

4.2.2.3 Ecofeminist quilting and countering dominance over Nature

Teaching, researching, and reading environmentalism through feminist intersectionalities is complex and difficult, and is endless in analysis. We return to the arguments in Chapter 1 that ecofeminism is sociohistorically plural in deconstruction beyond gender-focused oppressions, similar to ecopedagogical work that focuses upon not only environmental issues but also the intersectionalities of sociohistorical oppressions that lead to socio-environmental oppression and unsustainability. There are numerous definitions of ecofeminism, but Stephani Lahar's definition clearly expresses these aspects.

> Working out an integrated philosophy of humanity and Nature is not only to challenge dualisms to reflect more clearly our lived experience in theory but also to describe relations among women, men, society, and nonhuman Nature in a way that is conducive to a high quality of life and antithetical to oppression and exploitation . . . acknowledge[ing] and integrat[ing] rational, emotional, visceral, imaginative, and intuitive modes of experience and expression. (Lahar, 1991)

The plurality of ecofeminism leads to the difficult question of how does one form the theories' essence with intersectionalities between such disciplinary and epistemologically diverse theories and philosophies. I would argue this same question is also one for ecopedagogical work, as well as other critical work needing theoretical frameworks rather than a single theoretical lens. One important way is "quilting." Quilting of theories is explained by Karen Warren with the aspect theorizing, including ecofeminist theories and philosophy, as sociohistorically fluid.

> Theories are like quilts. The "necessary conditions" of a theory (say, ecofeminist philosophical theory) are like the border of a quilt: They delimit the boundary conditions of the theory without dictating beforehand what the interior (the design, the actual patterns) of the quilt does or must look like. The actual design of the quilt will emerge from the diversity of perspectives of quilters who contribute, over time, to the making of the quilt. Theory is not something static, preordained, or carved in stone; it is always theory-in-process . . . nothing that is knowingly, intentionally, or consciously naturist, sexist, racist, or classist-which reinforces or maintains "isms" of domination belongs on the quilt. Nor does anything that is not, in some way, about nonhuman Nature or human-Nature relationships. An ecofeminist philosophical result will be made up of different "patches," constructed by quilters in particular social, historical, and material contexts, which express some aspect of that quilter's perspective on women-other human Others-Nature interconnections. One cannot know, beforehand, what the actual interior of the quilt will look like. (Warren, 2000)

As Warren further explains, the patches can be quite different; however, the quilt becomes ecofeminism overall that counters domination and exploitation. With globalizations largely differing and diversifying (i.e., widening) paths toward understandings of feminism (Vargas, 2003), I would argue that such quilting is increasingly important.

For example, when globalizations have the endpoint of Westernization, quilting can help challenge "Western domination of 'women' and 'Nature' as conceptually linked and that the processes of inferiorisation have mutually reinforced each other" (Twine, 2001), and Western feminism often dominates the feminisms of the South (Warren, 2000; Alaimo, 2008). Within the field of critical literacies and global meaning making, Tierney discussed how quilting is essential toward global meaning making.

> Patchwork quilt explores readings of our worlds against the push and pull of internationalization versus Indigeneity, standardization versus ecological diversity. The metaphor of a quilt will resonate with those literacy educators wrestling with cross-cultural matters in an age of standardization of our educational and research practices and involved in interrogating their own

> ethics and practices in cross-cultural research and practices. As a quilt, it is a work in progress, with much to be added and stitching yet to be done. (Tierney, 2018b)

Within ecopedagogical literacies, such quilting is essential, in that the patches themselves coincide with ecopedagogical foundations as discussed in the previous sections, but that theories toward this goal cannot be narrowed in depth and width, and additional theories/philosophies need to be added. Playing off the wording of Tierney, there are always additional patches "with much to be added and stitching yet to be done."

In my own ecopedagogical work, I have utilized Warren's (2000) ecofeminist model which stresses the need for ecofeminist work to be inclusive of the following three components: "feminism (and all the issues feminism raises concerning women and other human Others); Nature (the natural environment), science (especially scientific ecology), development, and technology; and local or indigenous perspectives" (Warren, 2000). Warren uses a Venn diagram to illustrate that ecofeminism must be aligned with all three components together, and that the leaving out of any would result in it not being "ecofeminist" (Warren, 2000). Without question, all of these components must be problematized, I would argue within ecopedagogical analysis and reading that is presented within this book—for example, what is defining "development" and "science"?

Warren (1994, 2000) has expressed throughout her work that ecofeminism is not narrowly connecting feminism with ecological issues, but rather connecting the complexities of exploitation and domination within the world and Earth. Ecofeminist philosophy provides an essential logic that expresses the differentiating oppression of the world with the rest of Earth as only dominated, and the non-anthropocentric sphere being without abilities of choice and unable to resist environmental violence from the world.

> [Knowledges are] developed under domineering influence of males and other socially oppressive forces (Warren, 2000). Dominance differs from oppression, since oppression is limitation of choices. Nature does not have choices to be suppressed, but it can be unjustly abused (Warren, 2000). . . . This first type of ecofeminism calls for a "'cognitive dissonance' . . . to motivate one to re-examine one's basic beliefs, values, attitudes, and assumptions—one's conceptual framework," to dismantle thought processes of superiority of humans over Nature, and false vertical relationships between humans. (Warren, 2000) (Misiaszek, 2012)

As I discussed previously in this book, this oppression versus dominance can also be linked to subjectivity, reflectivity, histories, and dreaming of the world to determine acts of environmental violence as compared to the rest of the laws of Nature which tend toward survival and returning to balance.

Notes

1 The phrase "significantly destroy Earth" is used in the use of nuclear weapons to acknowledge the aspect that the rest of Earth will continue with or without humans, with adaption/evolution over time with significant damages that might return the non-anthropocentric sphere to exact sameness as we know it today.

2 Concept discussed by Massimiliano Tarozzi (University of Bologna, UCL IOE) at invited lectures at Beijing Normal University in October 2015 and in personal discussions at UNESCO meetings in Hanoi, Vietnam, in June 2019.

3 Ecopedagogical research that I conducted in Brazil, Argentina, and the United States found that one's environment is integrally connected to "cultures, traditions, values, languages and all else that makes up communities and that we cherish." Misiaszek (2011).

4 It is important to note again that oppressor-oppressed positionality is not fixed or binary, but rather according to the Freirean concept everyone is an oppressor or oppressed depending on the context.

5 Argument from Carlos Alberto Torres given during a lecture at Beijing Normal University, China, in 2015. For more information on ecopedagogical problematizing this argument see Misiaszek (2018b).

6 Boaventura de Sousa Santos De Sousa Santos (2018) has discussed the need for *sociologies of emergences* by countering *sociologies of absences*, in which only epistemologies of the North ground sociologies and define what is "legitimate" in the discipline. This needed paradigm shift of sociology radically reconstructs how the world is understood by countering and unlearning epistemologies of the North as apolitical and unquestionable as the only lens to view the world and Earth. See Chapter 5 for more in-depth discussions on this.

7 Some of the central theories I have utilized are the following: theories of globalizations, theories of coloniality, theories of feminism, critical race theories, media culture theories, various Indigenous theories, various theories of development and sustainable development, and dependency theories.

8 It should be noted that the separation of Freire from "Western-centric critical thinking" is very complicated and, I would argue with others Torres (2014), that Freire based much of his work on critical theories from the West with reinvention for his pedagogy and philosophy.

9 I have placed "nations-states" within quotation marks to signify the false bordering decision-making by the West to divide up continents to be colonized, as especially true for the continent of Africa Assié-Lumumba (2008).

10 I need to recognize that asking what parts of the "South Africa's population" can be problematic for many reasons, for example, colonizers as invaders are part of the population and the labeling of "South Africa" for the region is the colonizer's naming.

Part Three

Teaching Ecopedagogical Reading

5

Reading through Diverse Epistemologies and Methodologies

This chapter will focus on ecopedagogical work on problematizing epistemological frameworks through which we read environmental violence upon the world and Earth, as the world's epistemologies are diverse, most often conflictual, and have hierarchical issues between them, especially due to globalizations and coloniality. Epistemological discussions have already taken up this point; however, this chapter adds and extends to them, as well as analyzing the epistemological aspects of this book's key topics overall. This section will begin with the following overview of the key aspects of epistemological problematizing with ecopedagogical work and the three sections that follow will further define and exemplify the epistemologies of the South.

A key question to problematize in critical epistemological work within the world is what do knowledges and sciences "do" for people and populations, and how does it benefit some and not others. What must be first determined is which epistemologies are dominant and which aren't, and why. Answering this question calls for critical, rigorous sociohistorical analysis—for example, how has coloniality through globalizations from above brought about the fitting term of neocoloniality and what roles of dominant epistemologies have both helped create continuing coloniality and helped sustain/intensify it? For this important example coinciding with colonialism, modernity constructed from epistemologies of the North allow for "colonial appropriation and violence . . . [to be] converted into the colonial form of social regulation" (de Sousa Santos, 2018). Such oppressive regulation is reasoned through singular dominant ideologies that devalue all other ways of knowing—"immobilizing, dismissing, and demeaning the epistemologies and cultural practices of the South and people who are Indigenous" (Tierney, 2018b). These are just a few key aspects of continuing coloniality, more of which were given in the previous chapter, and will be given throughout this chapter and this book.

Questioning what knowledges do for "us" must also include the rest of Nature in the "us," which is also an epistemological question on what ways of knowing "allow" for this and what epistemologies do not? The first part of the question implies that most knowledges of the world are subjective, and the laws of Nature of the non-anthropocentric sphere are politically selected within the world. Within ecopedagogies, it is Freirean reading and rereading of how ways of thinking, including those emergent from sciences and from academia, coincide and contrast with ecopedagogical goals of balancing environmental violence for social justice with planetary sustainability. Paulo Freire's statement expresses the need to continuously read and reread the politics of sciences:

> As she [the teacher] underscores fundamental importance of science, the progressive educator must also emphasize to poor boys and girls, as well as to the rich, the duty we all have to permanently question ourselves about in whose favor . . . or in favor of what, we make science. (Freire, 2004)

One of the key aspects of how hegemony and environmental violence continue is through a singular epistemological framing without having other ways of knowing, which leads to problematizing the politics of S̲ciences that help sustain/intensify hegemony, socio-environmental oppressions, and planetary unsustainability. Such a Freirean reading coincides with Snively and Corsiglia (2001) problem-posing the following four questions on the problematizing s/Sciences (S̲ciences would be very similar to the term "Western Modern Science" (WMS) in the contexts of these questions):

(1) Is science an exclusive invention of the Europeans, or have scientific ways of thinking and viable bodies of science knowledge also emerged in other cultures?
(2) If WMS is taken to be universal, what is the status of the vast quantities of local knowledge that it subsumes, incorporates, and claims to legitimize?
(3) What is the proper role of science educators as leaders in the process of refining and clarifying the current definitions of WMS?
(4) When viable bodies of useful scientific knowledge emerge in other cultures, how can science educators develop programs that enable all students to cross-cultural borders—in this instance, between the culture of Western modern science and the cultures of the long-resident indigenous peoples?

The need for s̲ciences to critically counter S̲ciences is often epistemologically "dismissed as a nonproblem, given the broad-based consensus about what is and what is not science" (de Sousa Santos, 2018) with "knowing" that "global benchmarks . . . are exclusively Western and an acquiescence to Western standards of traditional empiricism" (Tierney, 2018b, 2018a; Zhao et al., 2017).

Ecopedagogical work deconstructs how Knowledge and Science systems frame the non-anthropocentric sphere, development, sustainability, citizenship, the rest of Nature, and many other aspects of environmentalism to counter them with through knowledges and sciences. This includes the need for decolonizing Knowledges and Sciences to determine how they continue with neocolonialization through globalizations from above that legitimize Development through "necessary" environmental violence. (De-)postcolonial scholars emphasize the need for multiple ways of knowing, in which de Sousa Santos (2018, 2012a) uses the term "ecologies of knowledges"—which means not only to learn through other epistemologies but also to "unlearn" knowing that is inherently oppressive, dominant, and unsustainable.

As Robert J. Tierney has cited my own work and those of others, he calls for multiple, diverse ways of knowing to teach literacy for global meaning making to understand how dominant epistemologies have led to environmental destruction, epistemicide, and widespread cultural extinction.

> Importance of differences would be considered essential (Grigorov and Fleuri, 2012; Misiaszek, 2015). Accordingly, global meaning makers should employ an ecological lens that questions the influence their research has on diverse cultures, especially the impact on our epistemologies over time—the extent to which our research builds upon and complements the local or traditional ways of knowing, or displaces or contributes to their extinction, what de Sousa Santos (de Sousa Santos, 2014, 2007a, 2007b) describes as their epistemicide. (Tierney, 2018b)

Widened learning, reading, and research calls for epistemologies of the South, including the pluralizing futures of development, modernization, sciences, and technology beyond confines of epistemologies of the North and histories of colonialization (de Sousa Santos, 2018). Ecopedagogies must include the aspects that Tierney has discussed in the quote on what is global meaning making, which is discussed throughout this book (as well as throughout his work). Within Freire's arguments that education needs to be utopic rather than fatalistic (Freire, 2004, 2000, 1992), widened epistemological perspectives are essential for any hope of ending socio-environmental violence and planetary sustainability. The epistemologies of the North, which are defined by de Sousa Santos (2018, 2012a) as inherently oppressive and dominant, cannot lead to such hope and would result in the goal becoming nonexistent. Thus, there is the overall argument that teaching through only epistemologies of the North will almost certainly result in socio-environmental violence and unsustainability.

Ecopedagogical work does not dismiss epistemologies but counters oppressive and dominant ways of knowing with other epistemological framing that are not oppressive or dominant. I agree with Graham Hingangaroa Smith's argument (cited in (Tierney, 2018b)) for the need to utilize any theories and practices that help with emancipation from Indigenous struggles. However, this does not dismiss the need for sociohistorical analysis of epistemologies, theories, and practices with their usage, as de Sousa Santos (2018) has called for, but I argue that there are possibilities for reinventing, as discussed in the previous chapter.

In addition, ecopedagogies counter epistemicide, as termed by de Sousa Santos (2018, 2016, 2007b), in which othered epistemologies are no longer seen as relevant. Inherent to the Freirean grounding of ecopedagogies, a socio-environmental oppression that is justified by dominant ways of knowing does not allow for reading of the world and Earth for necessary transformation. Rather socio-environmental transformation is blocked, as limit situation(s) for socio-environmental justice and sustainability, from a singular oppression-based epistemology that dismisses others' perspectives, with globalizations from above instilling singularity in the ways of knowing.

Environmental teaching through a single, dominant epistemology is also dehumanizing as it is pedagogically fatalistic. Such teaching does not allow reading and dialogue from multiple perspectives, and futures become a singular future with the goal of Development as legitimized through Sciences and Knowledges. Allowing only a single dominant epistemological perspective on how we define our development defies our humanness as diverse individuals who, with dialogue with others as social beings, have individual and collective histories and dreams. Singular-Knowledge epistemological systems prohibit reflecting through knowledges to understand and use sciences and have hope for possible futures—among other liberating/empowering reflectivity.

Beyond defining development, epistemologies of the North constrain the accompanying "sustainability" when only viewed through these dominant perspectives. There are various important questions on epistemological defining and measuring of sustainability, including what is the global "consensus" of the baseline for sustainability, a topic I problematized in Chapter 2. In other words, as a reminder, what is being "sustainable" and how is the baseline of what is sustainable epistemologically determined? Tamara Savelyeva (2017) argues the dominance of Western sustainability:

> Western sustainability as a whole is regarded as a hegemonic discourse, which often neglects those traditions that originated outside western borders . . . neglect of local and regional sustainability traditions by western scholars is perhaps the most critical entropy of the dominant western discourse and this trend can clearly be seen in education.

Within neoliberalism and Western individualism of the West, the sustainability baseline is largely determined by the Market (or the global Market in neoliberal globalization) with the gain of livelihood in of individuals' private sphere (will be discussed in more depth in the next chapter). With global inequalities, sustainability continues to not be constructed within globally all-inclusive socio-environmental justice or planetary sustainability. Sustainability is taught within epistemologies of the North as only to be prioritized when unsustainability affects one's private sphere of the powerful, without regard for much of the rest of the world or, even less, the rest of Nature. Epistemologies outside of those from the North, such as epistemologies of the South, must counter such oppressive-dominant perceptions.

5.1 Epistemologies of the South

It is important to begin with de Sousa Santos' (2018, 2012a) two key tenets of epistemologies of the South to help define them. First, epistemologies of the South exist to counter epistemologies of the North. Second, the "South" and the "North" are not geographic in their references but rather denote the oppressor and the oppressed, the dominator and the dominated, and the (neo)colonizer and the (neo)colonized, respectively, among many other unjust power dynamics such as de Sousa Santo's (2018, 2012a) inclusion of patriarchy and capitalism. Putting these two tenets together, the countering of the ways of knowing (i.e., epistemologies of the North) that justify the world's oppressions, dominance over Earth, and planetary unsustainability both defines and is the reason why epistemologies of the South exist. Regardless of geographic location, de Sousa Santos (2018) stated that epistemologies of the South are "produced wherever such struggles occur, in both the geographical North and the geographical South."

There are three specific characteristics of epistemologies of the South, which I will lay out here in this section, but it is far from an exhaustive list, with such a task needing another book (or, better, a series of books). First, epistemologies of the South are inseparable from oppressions, with "the knowledges that emerge from social and political struggles and cannot be separated from such struggles" (de Sousa Santos, 2018). There are various foci from scholars in which they categorize the differences between epistemologies that reproduce oppression versus those that counter them. For example, de Sousa Santos (2018) focuses on oppressions from capitalism, colonialism, and patriarchy, as opposed to epistemologies of the South in which the "production and validation of knowledges [are] anchored in the experiences of resistance of all those social groups that have systematically suffered injustice, oppression, and destruction." I would argue within ecopedagogical perspectives that dominance over Nature must be added to the foci on epistemological work to end socio-environmental oppressions and planetary unsustainability.

This is not to say that postcolonial, feminist work (among other critical foundational work) on epistemologies has not, and does not, already include such environmental aspects, but in ecopedagogical work these aspects are grounded on Freirean foundations for the analysis. World-Earth epistemologies are grounded on ecopedagogical work, in the same way that Gloria Ladson-Billings (2009) states there are many oppressive forces but critical race theories (CRT) centers on race as a sociohistorical oppression.

> [The CTR] movement used some of the tenets of CLS [(Critical Legal Studies)] but insisted on making race a central feature of their scholarly inquiry. By making race the primary feature of their scholarship and not just one among many, CRT scholars were making an important statement about the centrality of race.

Second, and as previously stated in this section's introduction, "there are epistemologies of the South only because, and to the extent that, there are epistemologies of the North" (de Sousa Santos, 2018). The former allows for understanding the suffering and domination caused by the latter, and without the former, suffering and domination continue without ways of knowing to counter it. It is also important to understand that the goal of using epistemologies of the South is not for them to be dominant epistemologies but rather to counter dominance from those of the North by forming multiple, diverse ways of knowing—that is, ecologies of knowledges as defined by de Sousa Santos (2018). This coincides with Tierney's (2018b) discussions on global meaning making in that using indigenous ways of knowing is "not displacing one epistemology with the other but a matter of positioning Indigenous knowledge as significant or primary, with the possibility of it being separate, fused, or integrated with Western ways of knowing."

That ecopedagogical work must be through multiple ecologies of knowledges coincides with Gloria Snively and John Corsiglia's (2001) arguments that we must understand the world and Earth through multiple and diverse sciences including, but not only, Western sciences.

> Cross-cultural science teachers will need a curriculum that recognizes a community's indigenous knowledge or worldview in a way that creates a need to know Western science (Cobern, 1994; Pomeroy, 1994). . . . Although the two perspectives may interpret the world differently, students should also see that the two overlap and can reinforce one another.

Snively and Corsiglia (2001) argue this need for environmental praxis through determining what is needed to solve environmental unsustainability with sciences that "also refer to conceptual constructs approved by logical empiricism (positivism) which, in addition, has the capacity to carry science beyond the realms of observation and experiment." Thus, they argue for the need to widen current understandings of sciences beyond and including Western sciences. As such, Western sciences would be utilized, and possibly reinvented, as Western sciences that do not dominate and devalue all other sciences—if they did, they would be written as Western Sciences.

Disruption of epistemologies of the North is essential within ecopedagogies—disrupting hegemonic ways of knowing that justify socio-environmental violence and planetary sustainability. Western ways of knowing have largely delegitimized Indigenous epistemologies and have largely narrowed our understandings of the world as part of the larger Earth. Tierney (2018b), with Indigenous scholars, has argued:

> Indigenous knowing represents a significant shift from Western reductionism and objectification to holistic reflections on the world that involve spiritual connections and a high degree of interrelatedness of people and their world.

With the focus of ecopedagogical work to deepen and widen our teaching, reading, and research (Misiaszek, 2018b), the narrowing of often false objectivity and reductivity fundamentally counters it. Also, widening our perspectives of the world-Earth cannot be accomplished within a singular epistemology but only through comparing and contrasting various ways of knowing. As stated repeatedly, it is all the worse if the *only* way of knowing is through Western epistemologies, so there should be calls for epistemologies that are non-Western.

Graham Hingangaroa Smith (2015, 2000) and Linda Smith's (2005, 1999) work has been cited by Tierney (2018b) as "cross-cultural engagements should proceed in a manner that is respectful of the histories, ways of knowing, needs, hopes, and values of all." Ecopedagogical uses of epistemological framings should be toward this goal, such as Western sciences but not in ways in which they become Western Sciences. Their arguments lead to the conclusion that all ways of knowing, practicing, and theorizing should be incorporated if such knowledges are to end oppressions; however, their analysis, with others, express how dominant epistemologies counter such widening of knowing. The importance of these arguments for ecopedagogies calls for me to quote Graham Hingangaroa Smith's (2000) argument again that I paraphrased earlier, that "we ought to be open to using any theory and practice with emancipatory relevance to our Indigenous struggle" (cited in (Tierney, 2018b)). I would argue, along with Smith later in his writing, that the use of Indigenous knowledges is essential for world-Earth struggles.

The use of Indigenous knowledges both allows for better understanding and offers possibilities of praxis of socio-environmental oppression within Indigenous communities, but also provides epistemological perspectives to praxis for the world-Earth holistically. Singh argues that Indigenous knowledges often allow for

> dispel[ing] "Nature-culture" duality, narratives of human mastery and exceptionalism and emphasize that other ways of being (other than as workers and consumers in the Capitalist desire machine) are not only possible, but already exist. They open up possibilities of other-than-capitalist subjectivities inspired by Indigenous ontologies and stance of humility, gratitude and connectedness with the rest of life on this shared planet. (Singh, 2019)

Without such epistemological inclusion, constructs of seeing the larger world as sharing the rest of Earth, rather than neoliberal goals of exploiting Nature, and the "others" of the world outside of one's Self, leads to unsustainable and untethered environmental violence and social injustices. Singh (2019) further uses Marisol de la Cadena's (2015) terminology of "anthropo-not-seen," coinciding with my ecopedagogical arguments throughout this book, for the "ongoing process of erasure of alternate worlds that do not make themselves through Nature-culture distinction and the impossibility of that destruction" (Singh, 2019). Fatalistic environmental pedagogies that reproduce world-Earth separations are promoted through a single epistemological lens, especially when they are epistemologies of the North. Ecopedagogies are aligned with Singh's arguments of the widening of environmental justice within degrowth framings (very similar to how I utilize *d*evelopment terminology) grounded on and deconstructed with epistemologies of the South.

Epistemological singularity and static-ness in teaching and reading environmental violence are oppressive at many different levels. First, there is the aspect that within diverse learning spaces, the teaching occurring is not meaningful to all those in the learning spaces. Second, the lack of epistemological diversity does not allow for those in the space to understand their own epistemological perspectives, in that perspectives from outside the proverbial fishbowl do not compare and contrast the Self's thematic universe epistemological groundings with those of others. Third, the Self's epistemological ideologies, including one's own "common sense," are left uncontradicted by others' ways of knowing, without understanding why others don't do "what I would do," thus only settling into "common sense" rigidity.

I will address this last one on "common sense" briefly, expanding on some of my arguments from Chapter 4. Within the larger global sphere, Carlos Alberto Torres (2011) argues that globalizations from above have continued to use neoliberalism and coinciding framings of "development" regardless of their failures.

> Neoliberalism has utterly failed as a viable model of economic development, yet the politics of culture associated with neoliberalism is still in force, becoming the new common sense shaping the role of government and education. This "common sense" has become an ideology playing a major role in constructing hegemony as moral and intellectual leadership in contemporary societies.

For consciousization to emerge on such deeply embedded "common sense" it is necessary to problematize what is most often not questioned, within other ways of knowing. With epistemologies of the North prompting neoliberal ideologies, countering ones from the South are essential. Torres (2011) further argued that there are three key questions that must be addressed, which are as follows:

> First, what is "common sense"? Second, was there a dominant common sense in the Western world that was modified by the advent of neoliberalism? Third, how can this new common sense can [*sic*] be defined in theoretical terms?

I would argue that to counter such neoliberalism as a tool for world-Earth oppressions and unsustainability, the epistemologies of the North that form such common sense must be countered by those from the South within ecopedagogies, as de Sousa Santos (2018) has argued overall. As I have argued elsewhere (2020, 2019a, 2018b, 2016a, 2015, 2011), including throughout this book, neoliberalism cannot coexist with global solidarity toward planetary sustainability—making common sense emergent from neoliberalism an enemy of environmental well-being, and thus a foe of ecopedagogues.

There are various essential aspects to the ecopedagogical goals of deconstructing neoliberal non-solidarity, which leads one to solely value one's own private sphere (á la Dirk Postma (2006), as previously described). One of the aspects is the widening of citizenships, which will be discussed in the next chapter (see Figure 1.1). Second, with the reinvention of such terminology, such as planetary sphere widening of the traditional framing of "citizenship," my previous research has found that such understandings must be understood through multiple epistemologies, especially epistemologies of the South.

Third and lastly, epistemologies of the South are grounded on understanding the limitations and the resulting incompleteness of "knowing." The incompleteness of epistemologies is acknowledged in epistemologies of the South that "take for granted that neither modern science nor any other way of knowing captures the inexhaustible experience and diversity of the world" (de Sousa Santos, 2018). This is opposite to epistemologies of the North that promote positivism with false objectivity. It is the diversity of knowledges that allows for better understanding of what are the limitations of the knowledges themselves and weeds out ignorance from them (de Sousa Santos, 2018, 2012a). As ecopedagogical goals include the need to understand limitations of knowing the world and the objectivity of the laws of Nature, as discussed previously, such understandings cannot emerge from a single epistemological framework, but needs diversity.

Epistemologies that counter epistemologies of the North, which promote the nonexistence of such limitations, are epistemologies of the South, as this is foundational to them. Also, there is the compounded aspect that epistemologies of the North reject other epistemologies, as opposed to those of the South in that, "given their pragmatic Nature, the epistemologies of the South do not, as a matter of principle, reject any form of knowledge" (de Sousa Santos, 2018). The utilization of diverse knowledges that counter hegemony and associated epistemologies of the North supporting hegemony is essential so that epistemological diversity is not viewed as a threat but is welcomed to end oppressions and dominance, as discussed previously in this section.

A key aspect of epistemologies of the South is that they emerge from experiences, especially experiences of those who are othered, which is essential in constructing ways of knowing to end oppressions. The need for bottom-up understanding is foundation in critical pedagogical work (Apple et al., 2009; Gadotti, 1996; Freire, 2000), as ecopedagogical work also includes understanding socio-environmental violence from those experiencing social violence from environmental violence. In addition to ecopedagogical work, epistemologies of the South deepen and widen understandings of how the rest of Nature, within the non-anthropocentric sphere, are dominated and destroyed by human acts of environmental violence. As epistemologies of the North limit knowledges within sociohistorical oppressions of colonialism, capitalism, and patriarchy, epistemologies of the South provide possible new ways of understanding what is the non-anthropocentric sphere's importance and its inherent worthiness outside of the anthropocentric sphere.

Within the three factors by which de Sousa Santo (2018) characterized epistemologies of the North, understanding beyond the anthropocentric sphere is impossible. In addition, leaving one's own foundation of epistemologies of the North, as one's "comfort zone," as de Sousa Santos discusses, or as I discussed Torres' (2011) work on "common sense," is epistemologically restricted.

> It also invites one to understand what one considers as relevant from the perspective of someone who does not consider it equally relevant. Leaving the comfort zone means to consider that what one does see and understand is part and parcel of a larger context that includes what one neither sees nor understands and what others do see and understand. (de Sousa Santos, 2018)

Essential here is that epistemologies of the South focus their knowledges from experiences with the question of how these first-hand experiences can be understood by everyone else for needed praxis. This is extremely complex and difficult but is foundational to critical pedagogical work (including ecopedagogical work) on how dialogue emerges within learning spaces outside epistemological framings <u>with</u> epistemologies within those learning spaces.

The initial need for such learning emerges from consciousization with the realization that there are other ways of knowing that are legitimate beyond those which "we" understand as "correct." In other words, it is necessary in consciousization to understand that what was previously learned as objective is actually subjective. De Sousa Santos (2018) discussed this aspect as follows:

> Dialogue among those experiences rather than forcefully imposing one of them upon all the others—makes no sense if one takes for granted that the objectivity of the world can be captured on the basis of one experience alone. If that were the case, one sole experience, however subjective and partial, could arrogate to itself the power to declare all others subjective and partial.

It is learning through multiple peoples'/societies' experiences that is necessary, as well as unlearning aspects that others' ways of thinking are not legitimate which, in turn, de-legitimizes others experiences within the world and Earth. All-inclusive social justice of the world is impossible without truly understanding one another, with environmental violence emerging from both such injustice and the inherent limitations of understanding the rest of Earth. Once again this coincides with the inseparability between social and environmental violence, and vice versa.

5.2 Disciplinarity and research of ecopedagogical work

This section delves into the complexities of academic disciplinarity of ecopedagogical work with connections to environmental pedagogical research, including epistemological analysis. The previous chapter had a section on the transdisciplinarity essence of ecopedagogies, while this section focused on how the epistemologies within and emergent from disciplines need problematizing that often leads to needing radical reinventions of them. The subsection will focus on how such research must be central within dialectic, problem-posing ecopedagogical learning spaces.

I will begin by problematizing the "hard" science disciplines, in particular with how the separation of "hard" Science disciplines from societies is inherently dehumanizing and deplanetarizing. This was initially discussed in the previous chapter's section on transdisciplinary needs within ecopedagogies. Specific to this chapter are discussions on needing transdisciplinarity to be also be grounded on epistemologies of the South. Freire discussed in depth that it is impossible to encase sciences and technologies, along with their education and education through them, into vacuums that are within the world (and I would argue the larger Earth) but separate from the world (and overall Earth).

> A type of curiosity that can defend us from the excess of a rationality that now inundates our highly technologized world. Which does not mean that we are to adopt a false humanist posture of denying the value of technology and science. On the contrary, it's a posture of balance that neither deifies nor demonizes technology. A posture that is from those who consider technology from a critically curious standpoint. (Freire, 1998b)

> The essential content in any educational program—whether on syntax, biology, physics, mathematics, or the social sciences—is that which makes possible discussions of the mutable Nature of natural reality, as well as of history, and which sees men and women as beings capable, not only of adapting to the world but above all of changing in it. It must view men and women as curious, engaged, talkative, creative beings. (Freire, 2004)

Dehumanizing, deplanetarizing Sciences and Technologies oppress by suppressing the reading of them within the world and without incorporation of the rest of Earth, beyond their manipulation (e.g., natural-resource mining and destruction of forest for corporate Farming) without consequences. With total separation beyond the already impossible apoliticization of sciences and technologies, deepened and widened understandings are absent. This does not only denies better understandings but also structures actions in the name of "science" to be done by limiting protest by suppressing knowing of associated oppressions and unsustainability. This is very much connected to environmental violence separated from social violence and planetary unsustainability due to Sciences and emergent Technologies.

De Sousa Santos (2018) deconstructs how disciplinary fields have led to domination through epistemologies and how determining to reinvent them is complicated. The key to problematizing disciplines, as de Sousa Santos argues, is deciding which epistemologies to use (i.e., those from the South or those from the North), where theories originate from and within what associated ideologies (i.e., for de Sousa Santos (2018, 2014, 2012a) the focus is on capitalism, patriarchy, and colonialization). Within the discipline of sociology, de Sousa Santos (2018) terms sociohistorical constructions from epistemologies of the North "sociologies of absences," which need to be radically changed to be grounded within epistemologies of the South, with the resulting discipline termed as "sociologies of emergences." This is a simplified description of de Sousa Santos' work, but an overall essential question within ecopedagogical spaces is what epistemological foundations are utilized in teaching, research, and reading. Within the three aspects of the North's epistemologies based on colonialism, capitalism, and patriarchy, world-Earth peace and sustainability are impossible (Misiaszek, 2020). The follow-up question is if transformative understandings and praxis can emerge through theoretical lenses to view the world-Earth and the disciplines (e.g., sociologies of absences) that are grounded upon such theories, "where" do new and reinventions of theories emerge from? Ecopedagogical work must include the deconstruction of the epistemological disciplinary foundations from which social and environmental injustice/violence have emerged and continue without being challenged within such foundations.

Within ecopedagogical work, such an epistemological paradigm shift from absences to emergences in teaching, reading, and research of environmental violence onto "us," "them," and the rest of Earth is essential as disciplines (in) directly justify resulting social injustice and unsustainability. Aligned with de Sousa Santos' terminology, "environmentalism of absences" must be critically deconstructed to determine the epistemological absences due to, and resulting in hegemony, for "environmentalism of emergences" to "emerge" for transformative praxis. Such problematizing of environmental studies/sciences disciplines seeks to answer what ways of knowing are structurally absent to better understand world-Earth connections that counter epistemologies of the North that utilize the disciplines, counterintuitively, for sustaining environmental violence rather than helping to end it. This coincides with my previous arguments of shallow environmental pedagogies helping to sustain/intensify environmental violence. I argue here the need to problematize the epistemological foundations of environmental disciplines (e.g., EE ("EE of emergences"), ESD ("ESD of emergences"), and ecopedagogy),[1] as well as disciplines that ground them (i.e., the social sciences, humanities, and the "hard" sciences).

As ecopedagogies are transdisciplinary, the emergences of the disciplines must follow such "___ of emergences" problematizing, including the "hard" sciences and their connections with other disciplines. As the objective laws of Nature are geared toward planetary sustainability, it is the politics of the world interpreting the interactions and the rest of Earth that must be analyzed for epistemological absences in the hard sciences, as "analysis of the conditions of identification and validation of knowledge in general, as well as justified belief . . . therefore, a normative dimension" (de Sousa Santos, 2018). The need for epistemologies of the South emergences in disciplines is expanded upon by de Sousa Santos as follows:

> Turning absent subjects into present subjects as the foremost condition for identifying and validating knowledges that may reinvent social emancipation and liberation (de Sousa Santos, 2014) . . . the epistemologies of the South necessarily invoke other ontologies (disclosing modes of being otherwise, those of the oppressed and silenced peoples, peoples that have been radically excluded from the dominant modes of being and knowing). (de Sousa Santos, 2018)

With the focus of listening to and amplifying voices that have been silenced by epistemologies of the North, emergences of disciplines are inherently countering oppressions. When disciplines silence ways of knowing, especially with knowing originating from historical socio-environmental oppression and Nature's domination, those ways of knowing, and the disciplines that originate and remain within them, are inherently oppressive, dominant, and unsustainable at their roots. Without the emergences of the disciplines of, including the foundations of, environmentalism, deepened and widened socio-environmental world-Earth understandings are barriers—or Freirean limit situations.

For the rest of this section I will focus on the research aspects of ecopedagogical work, although frequently returning to the disciplinarity arguments previously given. Critical pedagogical research, as helping to guide teaching that enables students to better understand the world and to structure education toward this goal, must be problematized on what theories, methodologies, disciplines, and epistemologies (and the intersectionalities between the four) to ground the research upon. Ecopedagogical research widens the "better understanding" goal to the planetary sphere, problematizing how these four either counter or reproduce socio-environmental violence and unsustainability, with these "reproducing" to be either removed or reinvented. Such reinvention would include the following epistemological arguments of epistemologies of the South and disciplines of emergences that de Sousa Santos argues (2018, 2016, 2012a, 2007b).

As I have written elsewhere (2018b, 2012, 2011), ecopedagogical research's grounding questions are as follows: How are the connections between social and environmental violence taught, not taught, and/or mistaught, and, most importantly, what are the politics behind these aspects of teaching? Ecopedagogical research focuses on the world's power dynamics on how education is reducing, sustaining, or intensifying how our environmental violent actions that result in social injustices and planetary unsustainability. There are many underlying dynamics toward answering these questions, including what epistemologies have countered or reinforced oppressive and unsustainable power dynamics of the world.

To put these arguments in the work of ESD, what is ESD research if not to inform us on how to better contextual teaching practices for <u>*d*</u>evelopment toward socio-environmental peace and planetary sustainability? In the following quote, Tierney discusses how scholars, like Gramsci, claim that academic research and teaching should be "organic" and carried out for transformative praxis, while also referencing similarly aligned works of Freire and Lather.

> Gramsci's [(1972)] view that scholars should be "organic" versus "traditional intellectuals," engaged with the public rather than remaining cocooned in and perpetuating only their academic theorizing and studies. This notion fits with interrogating the systems and is not unlike Freirean conscientização, or transformative change, or what Lather (1986) termed "catalytic validity." (Tierney, 2018b)

I would argue that "organic" research is essential for ecopedagogical reinvention of academic disciplines and research for socio-environmental justice and sustainability within all disciplines, as disciplinary planetarism is inseparable from humanism. This would return to the previous section's constructs on the need for epistemologies of the South to counter those of the North as widely held (i.e., global) legitimate knowledges. This would include epistemologies of the South grounding and reinventing disciplines, and thus needing to be "organic" through bottom-up research with "the role of researcher to that of a supporter and ally (Bishop, 2002) rather than distant observer, critic, or director" (Tierney, 2018b).

Such research must be bottom-up. Aligning with epistemologies of the South needing to be "with" the oppressed toward ending their struggles (Bishop, 2002; Tierney, 2018b), ecopedagogical research similarly focuses on "needing to be 'with' the oppressed" within world-Earth interactions and ways of knowing that position the rest of Nature and Earth holistically as the "the oppressed." Bottom-up research is essential for many reasons, with one returning to the question, what is the baseline of sustainability as we teach "sustainable development"—*d*evelopment or *D*evelopment (also see Chapters 2 and 6)? If research is guiding teaching through only practices grounded in epistemologies of the North, the result is ES*D* rather than ES*d*.

Intensifying globalizations calls for the rising need of widened global research and those identified as "global researchers" both complicates and allows for possibilities countering oppressions/domination. Toward this goal, Tierney (2018b), through the work of de Sousa Santos and Raewyn Connell, argues that teaching global meaning making must emerge from research that counters epistemologically dominant foundations which inhibit the use of "other" epistemologies.

> De Sousa Santos (2007b, 2007a) and Connell (2007) suggest that Western research has achieved a monopoly through protectionism that inhibits global knowledge developments, especially in non-Western countries. Insularity, nationalism, Western exclusivity, and control of global governance and outlets are part of the leveraging Western norms and conventions that preclude other epistemologies.

Within intensifying Westernization from globalizations from above, understanding outside of a singular Western epistemology (i.e., epistemologies of the North) becomes systematically more difficult, in that all other ways of knowing is abnormal, as well as delegitimizing non-Western-based research and teaching practices that emerge from it. Tierney (2018b) calls for global researchers to cross borders such that they are "not restricted to Western norms but that incorporate terms emanating from local cultural expectations, consistent with the notions of Giddens (1999), who argues for a global dialectic" (Tierney, 2018b). Calling for globalizations from below rather than those from above, Tierney ((2018b) citing (Takayama et al., 2017; Zhao et al., 2017; Chen, 2010; Connell et al., 2017; Park, 2017)) further calls for the need to problematize benchmarking within Western research methodologies.

If epistemologies of the North, rooted in what de Sousa Santos (2018, 2014) and others have argued, are rooted within capitalism, patriarchy, and colonialisms, with other scholars focusing on other sociohistorical oppressions, how do methodologies do anything but measure and form decision-making for sustainability within Development? To simplify a bit, if our measurements of sustainability are based on inequalities, supremacy of one over another, and domination of Nature, how can injustices and true planetary unsustainability not be the ultimate result?

Grounded on Freire's aspect of research as finding answers to ingenuous curiosities (Freire, 1998b)), and critical pedagogies grounded in praxis to help end injustices by understanding them and forming actions within bottom-up approaches (Gadotti, 1996), ecopedagogies center on understanding socio-environmental oppressions from those who suffer from them the most. Returning to the discussions on "common sense," Freire argues the need to problematize "common sense" within research as curiosities—for example, as "why" and what does such unquestioned knowledge do for "us"?

> To think correctly, in critical terms, is a requirement imposed by the rhythms of the gnostic circle on our curiosity, which, as it becomes more methodologically rigorous, progresses from ingenuity to what I have called "epistemological curiosity." Ingenuous curiosity, from which there results, without doubt, a certain kind of knowledge (even though not methodologically rigorous) is what characterizes "common sense" knowing. It is knowledge extracted from pure experience. To think correctly, from the teacher's point of view, implies respect for "common sense" knowing as it progresses from "common sense" to its higher stage. It also implies respect and stimulus for the creative capacity of the learner. (Freire, 1998b)

One of the key aspects of why ecopedagogical research is so important is that it helps to understand how ecopedagogical teaching and reading can disrupt epistemologies that construct "common sense" that socio-environmental violence and unsustainability is justifiable. Ecopedagogical teaching and research has the additional responsibility of seeking justice within both anthropocentric and planetary lenses which is inclusive of Earth as a whole. Research that supports, legitimizes, and/or is grounded in banking-education methods directly counters the tenets of ecopedagogical research. I would argue, as I do throughout this book, that socio-environmental oppressions and dominance of Nature not only resulted from banking, colonial education, but is grounded within its fiber. Such education, and the accompanying research, is essential for violence and unsustainability to continue, as well as the associated hegemony. Disrupting such "common sense" is difficult owing to various aspects, including not only its complexities but also various possibilities of fracturing its "no alternative" stronghold.

With research as emergent and partially guided by curiosities, we should approach sciences and technologies in the same way to determine what are the positive and negative aspects and for whom is it beneficial and for whom is it not. Focusing on ending socio-environmental violence and with goals of planetary sustainability, curiosity of how and what s/Sciences and t/Technologies should be utilized and frequently reinvented is essential. Freire emphasizes that reinvention of sciences and technologies "presupposes an openness that allows for the revision of conclusions; it recognizes not only the possibility of making a new choice or a new evaluation but also the right to do so" (Freire, 1998b). Freire discussed this further saying that positivism of "rationality" has also systematically suppressed our curiosities, especially through banking-education models.

> As a manifestation present in the vital experience, human curiosity has been historically and socially constructed and reconstructed. Precisely because the advancement of innocence toward criticalness does not take place automatically, one of the primordial tasks of progressive educational practice is exactly the development of critical curiosity never satisfied or docile. That is the curiosity with which we can defend ourselves from the "irrationality" resulting from certain excesses of our highly technological time's "rationality." This analysis, however, carries no falsely humanistic impetus against technology or science. On the contrary, it seeks to imbue technology with neither divine nor diabolic significance, but to look at it, or even observe it, in a critically curious manner. (Freire, 2004)

Rather than sciences as fixed Sciences, reinvention to transform the world with balance of the rest of Nature should be the goal. An example of technologies that are inherently dehumanizing and neocolonializing is weapons that retain and increase global power domination (Illich, 1983; Kahn, 2010; de Sousa Santos, 2018).

Returning to the essence of "popular" aspects of critical pedagogies, it is essential that academic disciplinary and research foundations and practices emerge from understanding oppressions from the persons who experience them. Within ecopedagogical disciplines and research, it is hearing from voices which suffer from socio-environmental violence, with the recognition of the limitations of truly knowing all the aspects of such suffering, that is essential.

Knowledges emergent from lived experiences are essential for this. One of the key characteristics of lived experience is that it is without choice and often without "ways out" ((Sarukkai, 2012) cited in (de Sousa Santos, 2018)), which should be reflected upon the limitations of outsiders' understandings of the othered.[2] This does not mean that understanding cannot exist, but there are limitations to them, as de Sousa Santos (2018) states in the quote below through differences between ownership and authorship.

> Sarukkai distinguishes between experience-ownership and experience-authorship: we are owners but not authors of our experience, just as we are owners of books without being their authors.

Within the aspect of environmental violence, we do, and will ultimately all, suffer from planetary unsustainability; however, this does not mean that we all suffer in the same ways or to the same degree. This aspect is expressed throughout this book in the aspects of how theories, if used correctly, can help outsiders to understand lived experiences.

Bottom-up theorizing allows for deepening and widening understandings. Bottom-up approaches involve not only theorizing but also construction and reinvention of the theory(ies) themselves through research; de Sousa Santos (2018) later pointed out that if it is only through the eyes of those who have had the lived experience (as he termed "thicker experiences" for epistemologies of the South to emerge), it would deny the theorizing of anyone other than ourselves (our society(ies)/population(s)).

> What enables intelligibility and transmissibility is translation. Translation allows one to relate a given experience to others, whether familiar or relatively unfamiliar.

However, there must always be the critical recognition that the greater the intensity of the oppression, the harder is the suffering of the oppressed groups, of those who do not have a choice but to suffer—"the solidarity and cooperation called for tend to be most scarce precisely when they are most needed in order to strengthen the struggles against oppression" (de Sousa Santos, 2018).

In the next chapter I will discuss more on planetary citizenship, in which Moacir Gadotti argues that Earth is a citizen—the most oppressed, voiceless citizen. If Earth is a citizen, Earth must be given a voice in research and, arguably, needs the most amplified voice. This voice is beyond the anthropocentric sphere but is heard within the anthropocentric sphere. Constructing research toward understanding how to amplify the voice of Earth is essential within ecopedagogies but is extremely difficult, complex, and needs diverse epistemologies. I am confident in saying that epistemologies of the North will not provide answers to this goal, but rather the opposite, emphasizing the need for epistemologies of the South.

Socio-environmental violence is unchallenged without problematizing why environmental violence is taking place. For environmental hegemony to sustain, "knowing" the links to social violence and other ways of knowing to value Earth must be suppressed. Such curiosities of why oppressions and dominance continue must be suppressed. Ecopedagogical research must help determine systematic, epistemological aspects within education (inside and outside of schools) that suppress such curiosities of why "we," as humans, commit unsustainable, violent environmental actions toward each other and the rest of Nature. Such research also questions our own disciplines and their epistemological foundations that support the violence and domination.

5.2.1 Research within ecopedagogical teaching spaces

> Once again, there is no such thing as teaching without research and research without teaching. One inhabits the body of the other. As I teach, I continue to search and re-search. I submit myself to questioning. I research because I notice things, take cognizance of them. And in so doing, I intervene. And intervening, I educate and educate myself. I do research so as to know what I do not yet know and to communicate and proclaim what I discover. (Freire, 1998b)

As Freire has indicated, without being incorporated into histories and past persons from one's own society not being "their history" and being devalued within the larger spheres of society, teaching and research is dehumanizing. Such problematizing in the previous section's ending happens not only within "formal research" in the field but also within learning spaces and through ecopedagogical reading. Due to the problem-posing essence of ecopedagogy:

> Ecopedagogical learning spaces make them spaces of research, with varying degrees of methodological rigor, including deconstructing how specific environmental pedagogies help or hinder development and sustainability pluralistic framings. In other words, ecopedagogical research is both for researching environmental pedagogies and teaching tools to deconstruct the politics of teaching socio-environmental issues. (Misiaszek, 2020)

Research within ecopedagogical spaces is Freirean-based, with the recognition that curiosity, that is the initial points and overall foundation of research, is ceaseless as Freire's quote beginning this subsection suggests with "continu[ing] to search and re-search."

Opposingly, banking environmental pedagogies dehumanize education by encouraging to end such curiosities, by systematically discouraging critical thinking, and limiting possibilities within dominant politics, including those of neoliberalism and Development. There are various reasons for this, but one is, in Freire's (2004) words, not allowing for their "intervention in the world":

> Human beings are by Nature inclined toward intervention in the world, as a result of which they make history. Therefore, they must leave in history their mark as subjects, and not the tracks of mere objects.

As discussed in the previous section at length, Freire described research as more methodological in finding answers to curiosities; however, that does not mean this rigor within methodologies and finding outcomes becomes only technical and thus void of humanization or, I would add, planetarization. Freire (2004) explained this as follows:

> None of that can be offered by the technicist or mechanistically understood education. It is important to underscore, for well-intended but misguided educators, that the more education becomes empty of dreams to fight for, the more the emptiness left: by those dreams become with technique until the moment comes when education becomes reduced to that. Then, education becomes pure training, it becomes pure transfer of content, it is almost like the training of animals, it is a mere exercise in adaptation to the world.

Ecopedagogical reinventing of what Freire is stating here, I would argue, would be the following: environmental pedagogical research must center and recenter research toward development and planetary sustainability goals.

I argue that most harmful environmental pedagogical research for the world-Earth is toward neoliberalist goals, in which the goals dictate the research methodologies and results to align with sustaining/intensifying economic power for a few elite. If research is aligned for education to become/remain a tool for hegemony (i.e., toward Development), research to inform technical teaching is best for students to become objectified objects as tools of their own oppression and planetary unsustainability. As Freire (2004) has stated, such technical education is equated to the "training of animals" outside of human's abilities for sociohistorical reflection (or utopic dreaming, as discussed previously); technical education serves the neoliberal goals of a few gaining from everyone else's oppressions and planetary unsustainability. In Freire's (1997) words, research (same as teaching), should not be fatalistic but rather utopic in its possibilities of guiding transformation.

> I feel serious work, meticulous research, and critical reflection about dominant power, which is gaining increasing dimensions, have never been as needed as they are today. The activity of progressive intellectuals must never equate that of people who, recognizing the strength of obstacles, consider them to be insurmountable. That would be a fatalistic position, alien to the task of the progressive. Understanding obstacles as challenges, the progressive must search for appropriate answers.

I should note that it is not so much that neoliberalist approaches want injustices and unsustainability, rather it is that there is no value placed upon social justice and sustainability, thus both are ignored or actively countered as obstacles in gaining economic profits and power. In addition, neoliberal approaches limit possibilities guided by emergent research within its own ideologies, thus making the research and "legitimate" methodologies fatalistic. As critical research, including ecopedagogical research, focuses on "dominant power" to determine possibilities of overcoming "obstacles" (i.e., limit situations), once again ecopedagogical work is named as an inherent enemy of neoliberalism.

Ecopedagogical practices and research inherently counter such overly technical pedagogies and research toward such environmental pedagogies. How can education, with guiding research, center on social justice and planetary sustainability? It is essential to problematize what this research looks like, with bottom-up construction and goals of better understanding the non-anthropocentric sphere. Such research is also accomplished within ecopedagogical spaces by determining what is necessary to know for praxis to end injustices and unsustainability and teaching false truths, as well its associated politics (the extreme of this within post-truthism will be discussed in the last chapter).

It is important to continuously remember that ecopedagogical problem-posing, inside and outside the learning spaces, is inclusive to the widest of spheres—the planetary sphere. Research within the sociohistorical contexts is essential but is too frequently ignored in research, especially those with endpoint goals of Development. Rather than research on society(ies)/population(s) that is not for bettering those society(ies)/population(s) that are researched, such as Development, educational research for social justice must be the goal. As such, ecopedagogical research focuses on planetary sustainability for social justice holistically throughout the world and on ending unsustainable violence upon Earth's non-anthropocentric sphere, beyond the anthropocentric sphere.

In the discussions throughout this chapter on epistemologies and methodologies that construct/legitimize disciplines, research, and popular understandings, there is the underlying need to read and reread how these intersecting structures either help sustain or help end socio-environmental violence and/or sustainability. For example, a key question is how Sciences' disciplines and methodologies position the world's oppressions and Earth's unsustainability. Ecopedagogical research within teaching problem-poses how Sciences are being used to justify such positioning, and Sciences can never justify oppressions and unsustainability, but rather they ignore humanism and planetarism as a falsely separated discipline. Freire has argued that sciences will never justify this, with the need to denounce sciences that attempt to do so.

> Nothing. The advance of science or technology cannot legitimate "class" and call it "order" so that a minority who holds power may use and squander the fruits of the earth while the vast majority are hard pressed even to survive and often justify their own misery as the will of God. I refuse to add my voice to that of the "peacemakers" who call upon the wretched of the earth to be resigned to their fate. (Freire, 1998b)

Ecopedagogical work must counter such justification, coinciding with Freire's refusal of such false "scientific" justifications (i.e., Scientific justifications). I would argue that this is the key aspect emerging from this chapter as a whole, as well as how we can reinvent disciplines, epistemological framings, and research to this end.

The next chapter will continue with this discussion in terms of problematizing "citizenship," which will call upon many of the arguments that I have given in this chapter.

Notes

1 I include ecopedagogy in this list of examples to emphasize the need for deconstruction of ecopedagogical practices to adhere to such disciplinary problematizing; however, ecopedagogy should have this as a foundational tenet already.

2 De Sousa Santos (2018) expresses the uniqueness of thick lived experience in the following quote:

> Nevertheless, resistance and struggle are not to be understood as exercises in freedom in the struggle to end oppression. Only those who do not live oppression in their flesh and blood could imagine that resisting oppression is a supreme act of freedom. Resisting and fighting oppression are often as much a necessity as living or experiencing oppression.

6

Reading through Citizenships

"Development," "Livelihood," and "Sustainability"

This chapter will connect aspects of solidarities grounding citizenship to ecopedagogical work to problematize aspects of d/Development, sustainability, and livelihood, because citizenship solidarity also means degrees of exclusion of those labeled "noncitizens." As discussed previously, citizenships in this book are used in the plural sense, with multiple spheres from local-to-planetary (see in Figure 1.1, the bottom two illustrations *Citizenship spheres (citizenships)* and *Citizenships complexities*). In this chapter, I continue my arguments that citizenship must be both widened and localized for education toward global socio-environmental justice and planetary sustainability, with each citizenship (i.e., "citizenship sphere") to be valued as a holistic "us" (i.e., "fellow citizens") within and between citizenships (Misiaszek, 2018b; Misiaszek and Misiaszek, 2016; Misiaszek, 2015, 2011). A key component of plural citizenships is intensifying globalization reaching most parts of the world and its effects upon Earth. The supra-global sphere of planetary citizenship is revisioning Earth as a citizen with the world as part of it within aspects of global citizenship.

> In planetary citizenship with Earth as a citizen, "an expression that was adopted to express a group of principles, values, attitudes and habits that reveal a new perception of Earth as a single community." (Gadotti, 2008c, 2008b, 2008a)

Citizenships form contested terrains within and between them of (dis) connecting social violence, environmental violence, and planetary sustainability. This chapter will analyze some of the main aspects of these contested terrains and offer possibilities for ecopedagogies for connectivity of all of the world within models of critical global citizenship and its education (GC/E) and all of Earth with planetary citizenship and its education (PC/E). A key question is how can we utilize theorizing and methodologies of comparative education, and ecopedagogies, within and between models of GCE, PCE, and ESD, are necessary to construct transformative pedagogies that teach the root political causes of environmental violence.

Although the term ecopedagogy was not around, it could be argued that its role in citizenship coincides with learning socio-environmental, sustainability knowledges that are essential in

> producing a citizenry that is knowledgeable concerning the biospherical environment and its associated problems, aware of how to help solve those problems, and motivated work towards their solution. (Stapp et al., 1969, p. 30)

To emphasize the following point again, it is us, as humans, who are the causes of this imbalance, as we are all responsible for planetary (un)sustainability, within the framing of social justice between ourselves and societies. This is from an individual to a global level, with the need for a global collective toward socio-environmental peace and sustainability, meaning that "global" must severe the "us verse them" roles of traditional citizenship, as well as planetary citizenship that includes Earth as a citizen.

We return to the aspect that humans' "development" is the cause for unsustainability, which counters Earth's equilibrium, within and between local-to-planetary spheres. In this book's terminology this would be *D*evelopment. However, our development can also lead toward sustainability, depending on whether its construction grounded or not in the ecopedagogical tenets discussed throughout this book. This would be *d*evelopment. The following are some key questions for citizenship education as connecting or not connecting to ecopedagogical tenets: Within education, what are the development ideologies taught and, equally important, not taught? How do differing models of "development" affect education, and vice versa? Does "development" include aspects of planetary sustainability and to what degree?

Livelihood, within this chapter, will be problematized as development within a more individual to smaller population spheres as they largely construct what is development, and vice versa. Through teaching by problematizing how different individuals and populations (including citizenship spheres) frame contextual *d*evelopment to increase their livelihood without negatively decreasing others' livelihood, ecopedagogical dialogue and reading on possibilities of development with all-inclusive livelihood can be achieved, along with local-to-planetary sustainability. This can be seen as a utopic goal. A key aspect of this is solidarity for one another and with the rest of Nature. Solidarity, as a core aspect of citizenship, and looking at the concept of livelihood through various spheres helps lead to an expanded view of progress and what should be sustained through multiple levels, from the local to the planetary. However, going toward this goal is also problematizing what is livelihood with the recognition that many current constructs of livelihood will never achieve this goal. Livelihood within neoliberal ideologies would be a key construct that is fatalistic in the sense that sustainability and socio-environmental peace will never be attained. Rather livelihood must be aligned with local-to-planetary solidarity, aligning with critical global citizenship and planetary citizenship. This will all be discussed further in this chapter later.

It is essential to problematized in ecopedagogies the constructs of citizenship that create exclusive or inclusive development models, as citizenship is a continuum between solidarity for "fellow citizens" (i.e., inclusion)along with the processes of exclusion between "us" and "them," the "citizen" and the "noncitizen" (i.e., exclusion). In addition, it is who can and cannot participate within and between citizenships and the thickness/thinness of that democratic participation that is crucial for ecopedagogical work. For example, who has a say in environmental violence inflicted upon them, from one's local society to our global society (e.g., from a leather tannery polluting a local river to a distant fossil fuel transnational corporation polluting numerous areas globally that more significantly affects global warming)? What does the inclusion of everyone within the society mean and the dynamics of democracy and its thickness/thinness is extremely complex? Within more local societies we have a better conception of this, but how does this relate to global citizenship democratic participation? I would argue that the contested terrain of globalizations provides barriers but also opportunities for thicker participation.

Thick democracy needs "the concept of a democratic citizenship where agents are responsible, able to participate, and choose their representatives and monitor their performance" (Torres, 1998b). Toward this goal, it is essential that we have critical pedagogies to deconstruct agents responsible for socio-environmental violence and then reconstruct toward a more socio-environmentally just world. Paulo Freire below discussed the need for education of citizenship to be "real," humanistic education full of dialogue for praxis to construct *d*evelopment.

> Fundamental background necessary for the full participation of any citizen in the development of the new society will be included in Basic Instruction. We are not talking about instruction in a school that simply prepares the learners for another school, but about a real education where the content is in a constant dialectical relation with the needs of the country. In this kind of education, knowledge, resulting in practical action, itself grows out of the unity between theory and practice. For this reason, it is not possible to divorce the process of learning from its own source within the lives of the learners themselves. (Freire, 1978)

However, how is this possible when some citizenship models label those as "noncitizens," sometimes with individuals/populations being a citizen of nowhere. Critical GC/E considers everyone as a citizen of the world; however, as discussed, other GC/E models help sustain, intensify, and/or widen those who are "noncitizens." Without needing to be stated the noncitizens have no or the very thinnest of democratic participation.

In ecopedagogical work, there is also the addition of planetary sustainability with such democratic citizenship, within and between local-to-planetary citizenships. Education for ecopedagogical literacy is essential, teaching students the tools to critically read all of these complexities (and endless more) and to then actively protest when local-to-global thick democratic participation is denied to them and for their democratic participation to emerge from ecopedagogical praxis. Problematizing in what ways citizenships either thicken or thin out our democratic participation and others' participation, especially populations who have been sociohistorically oppressed, is also an essential for ecopedagogical literacy.

Such thick democracy can be relatively understood up to the global sphere level, although incredibly complex. However, planetary citizenship cannot have a two-way democratic participation because only humans can participate through reflective decision-making. On this aspect, planetary citizenship is more conceptual but, none the less, important. As discussed with Moacir Gadotti in his office at the Paulo Freire Institute, São Paulo, over a decade ago and as described in much of his writing, if Earth is a citizen, within planetary citizenship education, it is the most oppressed, voiceless citizen. The rest of Earth does not "act" out of self-reflection, dreaming, and having histories, which are characteristics of being human as described by Freire (2000), or as John Dewey wrote that humans cannot be taught in the same ways as other animals (Dewey, 1963). In other words, acts emergent from justice can only be asked for from humans upon Earth (inclusive of humans)—reasoning that humans can only be oppressed while the rest of Nature can be, unfortunately, dominated by us. However, this does not mean that the rest of Earth does not have rights and value beyond what is placed upon it by human societies, which can be linked to theories of citizenship.

6.1 Problematizing citizenships: Local-to-planetary

To begin this section on problematizing citizenships it is important to discuss briefly the contested terrain of GC/E models, in which the following quote is from a book I wrote dedicated on ecopedagogy and GCE.

> GCE forms a *contested terrain* in the same way in which others have utilized the term for globalization, as I have discussed previously. As Torres pluralizes globalizations, and I have utilized the plural term of citizenships to signify the spheres of citizenship, I will also pluralize the acronym of GCE to GCE***s*** to signify its various models that both coincide and conflict with one another. The differences in the groundings of GCE models coincide with the processes of globalization, as contested terrains that can be either empowering or oppressive—"from above" (i.e., globalizer) or "from below" (i.e., globalized) (Torres, 2009; Stromquist, 2002b; Kellner, 2002; Torres, 2013a; Arnove, 2007). GC/E can be either contextually oppressive or empowering largely dependent on its local affects upon societies and populations (de Oliveira Andreotti, 2011b; Abdi, 2008; Abdi and Shultz, 2008; Richardson, 2008). To elaborate upon this topic, the significance of the denotation of contested terrains is that there are both positive and negative aspects which are not designated as one model of GCE as oppressive and another as empowering, but rather the need to deconstruct the goals and the politics of the goals of the models. In the actions to better understand the contextualization of global citizenship, deconstructing the politics of the power structures within and between societies and populations is essential in teaching and researching GCE (Marshall, 2011). Deconstructing the trans-national politics of power is essential to understand societies construction and legitimization of knowledges, ideologies, and loyalties. (Misiaszek, 2018b)

What I have detailed here on the contested terrain of GCEs through other scholars' work is very much aligned with previous discussion on the contested terrain of citizenships overall, in which this section will expand on briefly.

Although I will not go extensively into the arguments of GCE in this book, I would argue that a key ecopedagogical concern to deconstruct the contested terrain of GC/Es, as well as the contested terrain of citizenships overall, is their roles in local-to-planetary de-distancing or distancing. The term of "distancing" within the world can be connected to the aspect of citizenship and non-solidarity, in which epistemologies of the North has, in many ways, distanced the "us" versus "them" between "citizen" and "noncitizen," coinciding with many scholars' works (de Sousa Santos, 2018; Said, 1979; Memmi, 1991b). This gets more complicated when we discuss future citizens (see in Figure 1.1, the bottom illustration *Citizenships complexities*) (Misiaszek, 2020 (forthcoming), Misiaszek, 2018b). Without global solidarity, or without planetary solidarity with all of Earth, widened knowledges emergent from environmental pedagogical spaces are likely to be without ecopedagogical praxis as they lack the global connectivity needed for ending environmental violence globally. Boaventura de

Sousa Santos (2018) discusses the epistemological obstacles for active solidarity in the quote below.

> Active solidarity often implies effort and risk and in such cases only the clusters of similarity/proximity can activate ethical imperatives ("since this concerns me, I must get involved"). Conversely, the clusters of difference/distance are used socially to neutralize any possible disquiet from not being willing to run risks ("I don't have to get involved with something that doesn't concern me").

Constructing epistemologies of widened ethics, or the de-distancing of ethics, and overall caring, as I interpret de Sousa Santos' wording, needs to be done outside of epistemologies of the North and constructs of citizenship with them (e.g. neoliberal GC/E as defined by Lynette Shultz (2007)[1]). I would argue, coinciding with de Sousa Santos' overall arguments, that planetary sustainability is impossible within epistemological groundings that distance us from each other and from the rest of Nature (i.e., epistemologies of the North; e.g., sociologies of absences).

Ecopedagogical work on citizenships is impossible within such epistemological groundings, as well as impossible without countering suppressive epistemological groundings within education and research. De-distancing citizen models, such as critical GC/E and planetary citizenship that teach to question such knowledges are ecopedagogically essential, for praxis within de-distancing ethics and morality of the world as a whole and humans as part of the rest of Earth. Without needing to be stated, there are complex linkages between anthropocentric citizenships (local-to-global citizenships) and being *within* or *part of* planetary citizenship of "a new perception of Earth as a single community" (Gadotti, 2008c, 2008b, 2008a).

Ecopedagogues' goal of widening solidarity is teaching widened understandings of oppressions, violence, and exploitations, with the balance of respecting local ways of knowing. With the best of intensions of achieving this goal, the recognition of its difficulties in teaching needs critical reflection with solutions as utopic goals rather than perfectly achievable results. One is expressing oppressions from those who are most oppressed, which is expressed by de Sousa Santos (2018) as follows:

> [The] greater the intensity of the oppression, the harder it is for oppressed groups to communicate the suffering and the emotions that accompany their experience of oppression in such a way as to arouse active solidarity. . . . .harder the experience of oppression, the more difficult it is to share; in other words, the solidarity and cooperation called for tend to be most scarce precisely when they are most needed in order to strengthen the struggles against oppression.

De Sousa Santos explains that the solution for both the difficulties mentioned is for "us" to understand "their" suffering for necessary solidarity—solidarity through recognizing our limitations of "knowing" others' positionalities fully (see Chapters 4 and 5). Within my own wording on needing de-distancing for global and planetary solidarity, determining how ecopedagogues can better teach to read such difficult understandings of others' oppressions is essential, as well as reflectivity of the limitations. Once again, such teaching and methodologies of research must be focused to problematize the impossibilities of achieving this within epistemologies of the North, as their foundations counter such deepened understandings for solidarity.

For better widened reflection through deepened understandings, translation of oppressions and transformation from contextual, bottom-up approaches are possible, as de Sousa Santos (2018) continues his argument below:

> Given the constitutive intertwining of capitalism, colonialism, and patriarchy, liberation is possible only when isolation is superseded by reciprocity, solidarity, and cooperation among social groups that are differently but conjointly oppressed. At a time when there is so much diversity both in the experiences of oppression and in the struggle against oppression, intercultural and interpolitical translation is a precondition for such supersession.

Ecopedagogical work is to widen and deepen global understandings for widened solidarity to the global sphere, as well as determining what is necessary for better reflection upon limitations, along with the limitations of "knowing" the limitations. Such understandings are through experiences that are impossible to replicate for others fully;[2] however, top-down methodologies and theories/disciplines rooted in epistemologies of the North, without a doubt, distance such understandings (de Sousa Santos, 2018, 2014).

6.1.1 Economics and citizenships

> I cannot avoid a permanently critical attitude toward what I consider to be the scourge of neoliberalism, with its cynical fatalism and its inflexible negation of the right to dream differently, to dream of utopia. My abhorrence of neoliberalism helps to explain my legitimate anger when I speak of the injustices to which the ragpickers among humanity are condemned. It also explains my total lack of interest in any pretension of impartiality. I am not impartial or objective; not a fixed observer of facts and happenings. I never was able to be an adherent of the traits that falsely claim impartiality or objectivity. That did not prevent me, however, from holding always a rigorously ethical position. (Freire, 1998b)

Freire's passage above indicates the opposite of what citizenships should be, especially within the focus of critical global citizenship. The economic hegemonic focus of neoliberalism denies a large percentage of the world's population of humanism, which aligns with neoliberal global citizenship. I would argue that critical GCE is for students to problematize the world, toward praxis, through lenses that are "not impartial or objective" and "not a fixed observer of facts and happenings" to humans suffering under neoliberalism. For ecopedagogical work, this would center such suffering from environmental violence emergent from neoliberalism and widened planetarism for Earth holistically.

Aligning with the false "impartiality" or "objectivity" that Freire claims are characteristics of neoliberalism in fixing one's situation and their future, many of the aspects of neoliberalism discussed previously, such as emphasizing the singularity of a single fatalistic future without dreams of possible utopias to reflect upon plural futures, inherently counter solidarity with loyalty to only increasing one's own economic hegemonic positioning, and, as argued by Postma (2006), associating value only to the Self's private sphere—without solidary for others, especially "ragpickers." True solidarity of citizenships cannot exist within these and other characteristics of neoliberalism. Neoliberalism has the severity in which what is "cared" for ends up being the Self at worst and one's private sphere at best. Socio-environmental peace and planetary sustainability cannot succeed without valuing and having connectivity to widened spheres of the global and planetary.

Ecopedagogical deconstruction of globalizations and their effects on citizenship and livelihood are essential here in terms of economics. Through my previous research on expert ecopedagogues in Argentina (Buenos Aires, Córdoba), Brazil (São Paulo), and the United States (Appalachia region) (2015, 2011), the following was found through the data on citizenship, globalizations, and neoliberalism from the Argentinian and Brazilian participants.

> Participants expressed the changing constructions of citizenship with globalization from above and below, global neoliberalism emerged as a central concern as is affects citizenship. Participants felt that these influences ignored social and environmental elements of citizenship. The South's constructs of citizenship have increasingly mirrored citizenship constructs of the North through processes of globalization, with an emerging global citizenship that extends beyond nation-state borders (Suarez and Ramirez, 2007). Citizenship should be, "safeguarded from incursions by the market domain of buying and selling" (Marquand, 1991). Citizenship defined within neoliberal frameworks develops a citizenship with the same oppressive models that neoliberalism

> presents, especially in regards to gender, class and race. (Apple, 2006; Katz, 2008). (Misiaszek, 2011)

As globalizations have expanded the world, including economic possibilities, for some in the world, it also narrows "opportunities" for the vast masses, as neoliberalism is gain for a few with zero-sum ideology behind it. This is with hegemony sustaining rather than justice and sustainability. Outside the world, within the non-anthropocentric sphere, there is zero concern unless one's private sphere is economically affected negatively. This is one of the many central reasons why neoliberal GC/E is an inherent enemy of critical GC/E, as well placing these opposing aspects throughout sub-global citizenships. This need was found in this research also as many of the expert ecopedagogues "expressed that citizenship must allow for possibilities to better livelihood and the definition of livelihood must extend well beyond framing it within neoliberal ideologies" (Misiaszek, 2011).

The question is whom our fellow citizens are, including the systematic hidden neoliberal characteristics of economic gain for a few on the masses' backs. Unveiling these hidden truths is essential in environmental pedagogies on development—understanding which individual(s), group(s), society(ies) are benefiting and which are paying the price of environmentally harmful acts—inside economic analyses. Ecopedagogical work problematizes how constructs of citizenships can, with ecopedagogies, counter extreme individualism, and the single-valuing of the private sphere is a goal, to "care" for each other as part of Earth rather than possibilities of global-market economic exchanges for self-profit (Misiaszek, 2020, 2018b, 2016a).

This includes the previous arguments of needing to deconstruct livelihood, which often leads to the needs of ecopedagogical reinvention of livelihood, especially when the grounding is neoliberalism. Reflecting on the concepts of livelihood through the solidarity of multiple citizenship spheres helps to expand "our" progress with others, including "our" progress of livelihood as affecting Earth within the planetary sphere. As de Sousa Santos argues in the following quote, it is through isolation (e.g., neoliberal private sphere sole-valuing) that "reciprocity, solidarity, and cooperation" diminishes.

> Given the constitutive intertwining of capitalism, colonialism, and patriarchy, liberation is possible only when isolation is superseded by reciprocity, solidarity, and cooperation among social groups that are differently but conjointly oppressed. At a time when there is so much diversity both in the experiences of oppression and in the struggle against oppression, intercultural and interpolitical translation is a precondition for such supersession. (de Sousa Santos, 2018)

A key question is how can citizenships be reconstructed to deepen and widen while being grounded on humanization and planetarization rather than on unjust and unsustainable economics which neoliberalism espouses. Returning to and borrowing from the terminology of thickness/thinness of democracy, what is the thickness/thinness of citizenship with economics? How is citizenship(s) connected or not to economics? It is also necessary to problematize what type of economics, along with terms such as "development," "sustainability," and "livelihood"? Thin democracy of citizens is aligned with what Martin Carnoy (1997) argued was incomplete democracy, and thus incomplete citizenship, within neoliberal ideologies:

> The "incomplete" democratic politics of neoliberalism—a politics reduced to enhancing isolated individuals' solitary competitiveness in a Darwinian struggle. Freire's state is *constructive*, one where citizens are reintegrated through forming new political and social networks based both on information *and* critical analysis of their own situation in the global environment. Freire's state is also one of *solidarity*, including the marginalized, even when the global market has no room for them and exclusionary local ideologies segregate them. (italics by Carnoy)

Carnoy's arguments partially through Freirean scholarship are on State citizenship, but I would state that they are relevant throughout local-to-global citizenships with the recognition that the networks differ within and between the spheres. Ecopedagogical problematizing of *what* frames citizenships and *how* it does so must include deconstruction of the sociohistories of oppressions and dominance, globalizations, and thickness/thinness of democracy as it relates to economics (as well as social constructs outside of economics).

Considering all these aspects, an essential question to problem-pose is whether being a good consumer equates to being a good citizen. If yes, does poverty condemn people as, at best, "bad citizens" and, at worst, "noncitizens"? Within neoliberal framings, poverty becomes criminal and unpatriotic, within and outside national borders, with those from Weak States (i.e., the globalized, the Global South) being seen as leeches, rather than those from Central States (i.e., the globalizer, the Global North) being seen as oppressors and thieves of resources and labor—both over-generalizations being wrong.

I argue the need for ecopedagogies <u>with</u> citizenships grounded in socio-environmental justice to counter the commodification of Nature. Teaching for <u>D</u>evelopment and ES<u>D</u> models are without social justice or planetary sustainability, and simplify the world within systems of economic inequalities and the non-anthropocentric sphere to be commodified. Also, when livelihoods are measured and ranked within neoliberal ideologies, economic and power gain only defines increased "livelihood" rather than caring for one another and the rest of Nature. My previous research on ecopedagogues found them often focusing on how our consumption is increasingly tied to our citizenship, our development, and our livelihoods within neoliberalism's influence globally (Misiaszek, 2018b, 2016a, 2015, 2011). Rather, as argued, ecopedagogies must ground citizenships to counter citizenships based on neoliberalism.

6.2 Citizenship solidarity: Problematizing development and sustainability

Problematizing livelihood as both constructing and constructed from "development," with special focus on intensifying neoliberal influences, is essential in ecopedagogical spaces. Untethered accumulation and its associated power are inherently unjust and unsustainable, with ecopedagogical spaces teaching, learning, and reading how acts of environmental violence are connected to such oppressive ideologies wherein we don't care for one another or for sustainability of Earth. As a result, we have lost, or possibly many of us never have had, epistemologies of such widening solidarity that grounds citizenships.

Ecopedagogues problem-pose prioritization of the individual in development, especially the de-prioritizing of all supra-individual spheres. For example, in the classroom, they might ask the following: How is development defined and measured when prioritizing individual gain and devaluing society and the overall environment (i.e., planetary sphere)? Such questions are difficult because the private sphere is not singularly defined but rather extremely complex, with complicated and often contradictory connections with the public sphere. Ecopedagogical lessons on "development" must include discussions within and between private and public spheres, as well as a continuum of citizenship sphere from local-to-planetary. This includes juxtaposition problem-posing of livelihood as (dis)connected with framings of "development" (livelihood is discussed in a separate section later in this chapter).

Ecopedagogical problematizing of development and citizenship is constructed for the populations and societies in the world which, owing to the actions of humans, compete with one another rather than in favor of sustainability within the larger anthropocentric and planetary spheres. For example, nationalistic citizenship framings based on neoliberalism positioning environmental actions that conflict with <u>D</u>evelopment as (in)directly decrease the power of the nation-states, which makes these actions unpatriotic as they might affect the nation-state's supranational positioning. Problematizing in ecopedagogical spaces would include questioning why global positioning is not linked to socio-environmental justice and sustainability, and determining how we can critically reimagine citizenships toward this goal. Countering such nationalistic citizenships that encourage environmental violence in the name of *D*evelopment, on which neoliberal nation-state citizenship relies, is essential in ecopedagogical work.

The historical connections between *D*evelopment and citizenship are abhorrent. *D*evelopment largely emerged from histories of colonization, with the colonized non-citizenship having no *d*evelopment and being denied all possibilities of a future. Global *D*evelopment emerged from coloniality by laying out the framework for globalizations from above to continue and intensify, and hence is more appropriately named neocoloniality. With colonialization removing the constructs of "citizenship," the colonialized become foreigners in their own homelands—noncitizens who are unable to migrate from their homelands (Fanon, 1967, 1963; Memmi, 1991b; Said, 1979). With this came *d*evelopment/progress that was not their own; the colonized lost their livelihood for the benefit of the colonizers, and their identity as "humans" was erased along with their histories and epistemologies. Sustainability, if it mattered, was only measured against how it affected the colonizers With hope and utopic dreaming denied, colonialization takes away citizens' right to democratic, self-determined futures as endpoints of *d*evelopment, while safeguarding the colonizer's future (i.e., non-plural) through *D*evelopment. It also includes problematizing epistemological limitations of understanding the "development" and "sustainability" that emerge from colonialization, which makes the masses noncitizens, as well as how livelihood "should be" according to the (neo)colonializers.

Decolonizing citizenship education includes questioning how we understand being "civil" as we define development. Such deconstruction includes problematizing what it means to be "cultured" by asking through which perspectives we must understand it. Among other decolonizing lenses, work on Orientalism (Said, 1979) and epistemologies of the South/North (de Sousa Santos, 2018, 2016, 2012a, 2007a) are essential. The following quote by Freire explains how colonial practices had taken from the colonized by denying them of their histories:

> Only the colonizers "have a history," since the history of the colonized is presumed to have begun with the civilizing presence of the colonizers. Only the colonizers "have" culture, art and language and are civilized national citizens

> of the world which "saves" others. The colonized lacked a history before the "blessed" efforts of the colonizers. The colonized are uncultured and "barbarian natives." (Freire, 1978)

For ecopedagogical work within this section's topics, with regard to Freire's quote, it is necessary to problematize how (neo)coloniality continues <u>D</u>evelopment and dominant Northern citizenship (i.e., emergent from selectively false legitimization of "civilizations") by continuing to deny histories and civilizations that are foundational to constructs of development. With false perspectives of the "uncultured," how has (neo)colonialism removed histories and, in turn, suppressed possible futures in the name of "development" along with the (neo) colonized as noncitizens?

Inherently contextual, determining <u>*d*</u>evelopment that is globally inclusive but locally contextual is extremely (maybe limitlessly) complex; however, this does not mean there are no commonalities between its goals. If critical GC/E is inherently <u>*d*</u>evelopment that is globally inclusive and planetarily sustainable, the very goals toward this utopic end coincide.

6.2.1 Reimagining globalizations with development and citizenship

Previous sections have discussed ecopedagogical problematizing on how to disrupt globalizations aligned with neocoloniality; this section focuses on problematizing how (de)reconstructing citizenships can disrupt the neocoloniality with oppressive "noncitizen status" from globalizations from above. It is argued by many educational scholars that public education was created to strengthen citizenship within nation-states; however, the fluid Nature of citizenship is complicated through the processes of globalization by significantly widening the definitions of citizenship and its scope (Lima, 2007; Suarez and Ramirez, 2007). However, widening here in framing means widening in terms of not only inclusion but also exclusion. As was discussed previously, such citizenship education is often viewed as one of placing the othered in "their places," coinciding with sociohistorical oppressions; globalizations have possibilities to counter such oppressions, but also processes that sustain/intensify oppressions. For example, that social injustice emerges from environmental violence should be of concern but is often ignored.

This coincides with ecoracism scholarship that problematizes which communities become sites of environmental violence as those of the othered (e.g., communities of color) (Merchant, 2003; Melosi, 1995); globalizations from above distance socio-environmental violence at the global sphere. There are various examples of such global distancing, such as e-waste from technologies in the Global North ending up in the Global South to be recycled by the globally othered, which results in much higher rates of cancer, among other serious health issues; forced labor at low wages; and poor school attendance with many school-age children doing this job. As an important side note, this is one of many contradictions of environmentalism in which an environmental task of recycling short-age technological products becomes a much larger socio-environmental-justice issue through deepened and widened (i.e., global) ecopedagogical readings. This unfortunate conundrum is a good example for the need of widened and deepened ecopedagogical reading.

It is essential to emphasize again that globalization from below does not simply mean that whatever is from the local is empowering directly opposing what is constructed from the global from above is oppressive, and the former or the later is not necessarily globalization from below or from above (Olmos and Torres, 2009; Torres, 2003). In other words, the origins of the processes of globalization without contextual analysis do not indicate "from below" or "from above" outcomes. In other words, not everything from the local is good and not everything from the global is bad, nor does the intention of being environmental locally always indicate global and planetary socio-environmental justice or sustainability. The previous example of e-waste illustrates that not all environmental actions bring about justice that is globally peaceful. Without global-wide teachings, readings, and research of socio-environmental actions, e-waste recycling technologies could remain being seen as largely unproblematic for sustainability. Ecopedagogical praxis emerges from questioning how such much-needed recycling can be accomplished while ensuring planetary sustainability within global justice frameworks. Within the realm of citizenships, how can citizenship framings of solidarity help for global de-distancing with all environmental actions, including e-waste recycling?

We return to Freire's arguments that all humans need to be taught as subjects of the world, in that their histories and cultures should be valued as essential, with ecopedagogy opposing (neo)colonial objectification and with "authentic development" seen as bottom-up and as needing to be aligned with planetary sustainability. As globalizations intensify, education must align its processes so as to help toward this goal, with solidarity (e.g., citizenships) for all humans, societies, and Earth overall. Robert J. Tierney (2018b) discussed the need to incorporate global meaning making as our human instinct for our "shared interest in sustainability." This would coincide with Carlos Alberto Torres' (2017; 2016) arguments of the global commons of us all wanting to live in an environmentally healthy environment. Ecopedagogy questions whether our benefits from environmental violence are in balance with those of everyone else and whether planetary sustainability is maintained. For example, coal provides energy with the result of pollution in many forms, so how does this benefit some population(s) but negatively affect others locally to globally? This leads to various questions about the locations of coal mining, from which coal is often transported worldwide, and leads us to critically problem-pose whether there are other ways of producing electricity.

I have argued for critical GCE with ecopedagogies as essential for progress toward this goal, and as a defining factor of what is "development" (2020a,b, 2019a, 2018b, 2016a). Tierney discusses how global meaning making must counter globalizations characterized as neocolonialism, with ecopedagogical work needed to reinvent these globalizations, characterized with the phrase used throughout this book, globalization from below.

> Global meaning making is the view that all of us are planetary dwellers with a shared interest in sustainability. Therefore, we are interested in spaces that support cross-cultural engagements to which all can bring cultural capital that is distinctive to their meaning making, and all of us can actively contribute to a global trajectory built upon support for diversity where differences without violent intent are respected. (Tierney, 2018b)

Problematizing what is meant by global meaning making, in which we see ourselves as "planetary dwellers," is very essential, in addition to problematizing what our role is as a planetary citizen inside and outside the constructs of critical global citizenship.

Aligning with Torres' (2017, 2016) argument that global commons essential for successful critical global citizenship, Freire (1998b) had argued with a "universal human ethic" that allows for commonalities for understanding, respecting, and loving (Freirean deepened respect and connections) among the diverse human race. Fatalistic education models do not offer this possibility. As shown in the following quote from a work by Antonia Darder (1998), this "armed love" of Freire

> could be lively, forceful, and inspiring, while at the same time, critical, challenging, and insistent . . . [standing] in direct opposition to the insipid "generosity" of teachers or administrators who would blindly adhere to a system

> of schooling that fundamentally transgresses every principle of cultural and economic democracy.

Widened citizenships (i.e., to the planetary sphere) with better understanding and respecting narrowed citizenships (i.e., to the local sphere) is essential throughout education. How can environmental pedagogies teach to read the distancing of globalizations to reinvent the process to truly de-distance our socio-environmental effects so that environmental violence on "them" is also upon "us." This is with the hope that the term "them," as one of separation for socio-environmental caring, is abolished. In addition, there is the need for de-distancing with the rest of Nature, in which planetary citizenship with critical global citizenship provides possibilities toward this goal.

The incorporation of what Freire described as the "human ethic" is essential in grounding critical GCE goals and also expanding this human ethic to the planetary sphere within constructs of planetary citizenship.

> When I speak of a universal human ethic, however, I am speaking of something absolutely indispensable for human living and human social intercourse. In making this statement, I am aware of the critical voices of those who, because they do not know where I am coming from, consider me ingenuous and idealistic. In truth, I speak of a universal human ethic in the same way I speak of humanity's ontological vocation, which calls us out of and beyond ourselves. Or as I speak of our being as something constructed socially and historically and not there simply a priori. (Freire, 1998b)

This is in opposition to (neo)coloniality that only, if at all, speaks of ethics for "us" and not for "them." This is not only opposite of but also directly countering human ethics that "speak of humanity's ontological vocation, which calls us out of and beyond ourselves" (Freire, 1998b). Included with this is the need to counter global neoliberalism and to align *D*evelopment as connected to ethics or humanization, as Freire wrote in the passage below.

> The place upon which a new rebellion should be built is not the ethics of the marketplace with its crass insensitivity to the voice of genuine humanity but the ethics of universal human aspiration. The ethics of human solidarity. (Freire, 1998b)

With ethics very much connected to the constructs of citizenship globally, ecopedagogical work problematizes who is included when we speak of ethics, with the inclusion of "who" within the non-anthropocentric sphere.

Such universal ethics, including the connections to citizenships, cannot be taught within education that is fatalistic (e.g., banking environmental pedagogies). As ecopedagogical teaching is inherently utopic, development should be full of hope for increased humanism with planetary sustainability, thus endlessly countering a singular future of unsustainability and unceasing socio-environmental violence (e.g., neoliberal global citizenship, sub-global neoliberal citizenships). Included in this is the hope of bettering livelihood, but not at the expense of others' social injustices and planetary sustainability—in essence, countering the betterment of "our" livelihood by diminishing "their" livelihood. Within neoliberalism and neoliberal global citizenship, this is a concern because, as Postma (2006) has argued, only livelihood within "our" private sphere matters.

Complexities also arise with the Self having multiple identities that both coincide and conflict with one another (Olmos and Torres, 2009; Torres, 1998a).[3] Globalizations have largely increased the Self's number of identities, making this even more complex. As previously discussed, Torres argued that fulfilling personhood through individuals' diverse identities is central to citizenship education in an increasingly globalized world—naming it one of his three global commons (2017, 2016). There are also coinciding/conflicting complexities of identities in a person's citizenship, especially between traditional national citizenship, other citizenships, and planetary sustainability (both within and outside planetary citizenship constructs). It is ecopedagogical teaching and reading through the global inclusiveness of critical GCE and/or planetary inclusiveness of PCE that is essential.

The valuing of one's history as well as the diversity of histories is essential when talking about more traditional models of citizenship of the former and widened supranational critical citizenship models of the latter. With regard to the latter, it involves deepening understandings of others' histories that were devalued through neocoloniality and the widened understandings of planetary sustainability. To this end, it involves problematizing how we can transform, sometimes radically, globalizations for such deepened and widened understanding of ourselves, and of our environmentally violent actions as affecting others, through sociohistorical analysis, coupled with better understandings of Nature's laws.

"Who" is oppressed by environmental violence in the name of "development" is essential in ecopedagogical work, with the recognition that asking this question is almost always absent. Deconstructing what is development within the world includes centering problematizing on what is seen as *d*evelopment as they call upon others for their own *D*evelopment. In other words, for those who profit from *D*evelopment, would they find it just if the *D*evelopmental actions caused de-*d*evelopment within their private sphere? Freire states this clearly:

> They demand of the rest of the world now what they were unwilling to demand of themselves. One of the tricks of their fatalistic ideology is the capacity to convince submissive economies (which will be engulfed in this process) that the real world is this way, that there is nothing to be done about it except to follow the natural order of the facts. It passes off this ideology as natural or almost natural. It does not want us to see and understand the phenomenon as a product of historical development. (Freire, 1998b)

Ecopedagogues focus this question within the planetary sphere with regard to the demands of *D*evelopment upon the planet, often leading to unsustainability connected to social injustices (e.g., de-*d*evelopment) and devastation within the non-anthropocentric sphere. Teaching *D*evelopment is successful because of the ways Freire described above, as ahistorical teaching of ideologies that normalize oppressions of *D*evelopment. Coinciding with what Freire later described in some of his last publications (1992, 2004), *D*evelopment as defined in the book is also only possible by non-liberating education that intentionally justifies the destruction of Nature.

Returning to ahistorical teaching of *D*evelopment, the taking away of history does not allow for *d*evelopment and is inherently dehumanizing. There are many ways to discuss this aspect, with a central one being coloniality and globalizations that align with neocoloniality. I quote Albert Memmi's argument that the "most serious blow suffered by the colonized is being removed from history and from the community" (Memmi, 1991a), as he further discussed how the removal from history continues with their de-*d*evelopment.

> We have no idea what the colonized would have been without colonization, but we certainly see what has happened as a result of it . . . subdue and exploit, the colonizer pushed the colonized out of the historical and social, cultural and technical current. (Memmi, 1991a)

The politics of global powers have emerged from histories of colonial oppressions which continue with globalizations from above (i.e., neoliberal globalization and neocolonialism). Freire has argued that our curricula, especially hidden curricula, "pay[s] very little attention in the historical development of our society" (Freire, 1993). Without teaching these histories of socio-environmental violence and unsustainability as creating our present and forming our futures, we are left without the knowledge and epistemological tools to construct *d*evelopment.

According to Freire (2000), teaching *d*evelopment involves understanding that we are unfinished beings constructing unfinished societies "scientific[ally]" to develop through praxis, for deconstructing sociohistorical oppressions in situating populations in the world, as articulated by Stanley Aronowitz in his introduction to *Pedagogy of Freedom*.

> Freire stands firmly in the tradition for which the definition of science is critical and not positivistic. Educational formation becomes "scientific" when the learner grasps the link between theory arid practice through a process whose assumption is that the individual is, in every respect, "unfinished." The accomplishment of critical consciousness consists in the first place in the learner's capacity to situate herself in her own historicity, for example, to grasp the class, race, and sexual aspects of education and social formation and to understand the complexity of the relations that have produced this situation. Such an accomplishment entails a critical examination of received wisdom, not as a storehouse of eternal truths but as itself situated in its own historicity. Implicit in this process is the concept that each of us embodies universality but that it does not necessarily dominate us. Thus, the active knower, not the mind as a repository of "information," is the goal of education. (Aronowitz, 1998)

I would argue that teaching to suppress understanding situating of one's Self sociohistorically allows for *D*evelopment to be normalized, as well as situating humans to justify the domination of the rest of Nature. Knowledges, sciences, and technologies must be problematized, as Freire has discussed, through aspects of unfinishedness and historicity, rather than false positivism constructing *D*evelopment goals. It is important to clarify again that Freire did not underestimate the need for sciences but rather held the view that "hard" sciences are not the only disciplines to understand the world. He emphasized the need to teach to humanize sciences and emergent technologies rather than blindly following them as apolitical or "aseptic" (Freire, 1992), as discussed next. *D*evelopment teaching, as discussed previously, is for a single future based on neoliberalism, which, with its processes, sustains current hegemony, without historical context or manipulating ideologies based on false histories.

As a worrisome example, this manipulation aspect will be discussed in more depth in the next chapter under the topic of the current era of post-truthism. Post-truthism will be expressed as utterances that have the sole purpose of betraying truths and lead to, as Michael Peters (2017) has discussed, governance by lying. This is not a perfect or binary separation; however, this separation between the politics of constructing "truth" and post-truthism's construction of purposeful false truth, as understood by its creators, is how the two terminologies has been utilized in this book. Post-truthism also has called for Torres, UNESCO chair of global learning and global citizenship education, to possibly add a fourth global common to restore faith within sciences and knowledges.

Banking education does not allow for transformation because historical reflection is devalued to nil and teachers as depositing knowledge, with teachers' deskilled role to being "reduc[ed] to the imparting of content that would not even need to be treated aseptically, and aseptically 'transmitted,' since, as the content of a neutral science, it would already be aseptic" (Freire, 1992). The neutralizing of knowledge—of being "aseptic" for both the knowledges and the teaching—is inherently false by making both apolitical. Gadotti and Torres (2009) have discussed this at length within the contexts of Brazil not being able to be democratically developed largely due to its history of banking-education models.

As previously discussed, banking education systematically suppresses reflection that allows for decision-making for true development, but rather teaching for alienating Development as Freire (2000) discussed in other terms below.

> Men who are submitted to concrete conditions of oppression in which they become alienated "beings for another" of the false "being for himself" on whom they depend, are not able to develop authentically. Deprived of their own power of decision, which is located in the oppressor, they follow the prescriptions of the latter. The oppressed only begin to develop when, surmounting the contradiction in which they are caught, they become "beings for themselves." (Freire, 2000)

Teaching Development to students is not for them to become "beings for themselves" but rather to become tools of hegemony, and thus working, and various ways existing, for Development as falsely portrayed as development. In Hegelian terms within ecopedagogical framings, students become slaves for their masters' Development, with liberation and planetary sustainability only being able to come about once students recognize their power in Development processes and recognize it is not their own development—and that they are being "hosts" for others, as "noncitizens" and "non-global citizens." As previously stated, deep historical analysis of continuing socio-environmental oppressions problem-poses the following questions: How has their denial of participating in history ill affected the current environment? How has their removal from history made them the "false cause" of socio-environmental problems?

6.3 Citizenship solidarity: Problematizing livelihood

Returning in this section to livelihood as relating to development, I will discuss how ecopedagogical work problematizes the various ways how citizenships both narrow and widen gaps of solidarity between each other's' livelihoods globally and the resulting (dis)alignment with planetary sustainability. Livelihoods also form contested terrains that coincide and contrast with one another. These contested terrains need to be problematized to determine the inequalities between livelihoods that may decrease and/or devalue others' livelihoods, as well as determining the (un)sustainability aspects of them. Decreasing and devaluing livelihood can often be very extreme, affecting not just the livelihood but the mere survival needs of others, as discussed previously with Maslow's Hierarchy of Needs pyramid. The sustainability aspect is not questioning so much an individual's or often even a whole population's framing of livelihood but questioning sustainability if very large population(s) act toward achieving the livelihood in question. An extreme but necessary example is if the actions for sustaining/increasing the livelihood was followed by more than seven billion people. If ecopedagogues place importance in problematizing contextual livelihoods as globally all inclusive and within the limits of planetary sustainability, a key question is what framings of citizenships with global solidarity with one another as humans and planetary solidarity with the rest of Earth is needed. One of the difficulties here is that it is not only bettering persons/populations' livelihoods but also limiting and lowering livelihood when framed as unsustainable or unjust. For example, such lowering is likely to apply to much of the Global North. I don't have an answer to this and there might not be an answer, but this should be part of ecopedagogical praxis.

As ecopedagogues center the problem-posing questions of "what is," "for whom," and "who defines" on how we teach what is "development" and its "sustainability," the following coinciding questions also emerge: What is livelihood? Whose livelihood(s) "matter"? How does livelihood become "common sense" to strive to achieve? Ecopedagogical teaching through problematizing development and sustainable development must also include problematizing the more local contextualization of individuals' and societies' livelihood, as all this coincides and conflicts with ecopedagogical tenets. One aspect of livelihood that must be noted again is the differences between Western individualism that can be modeled with neoliberalism or not, and personhood as discussed earlier in this chapter. Being able to live through one's own identities (i.e., personhood) is essential for livelihood and *d*evelopment; however, this is can be much different from lifestyles of those from above. Coinciding with citizenship, how does citizenship affect how these questions are answered on livelihood and the ideological conflicts between citizenships?

I interpret Freire's passage below as introducing aspects of citizenship as "collectivity" with development and the aspects of one Self's and their societies' livelihood of labor.

> If production is governed by the well-being of the total society, rather than by the capitalist, private or state, then the accumulation of capital—indispensable to development—has a totally different significance and goal. The part of the accumulated capital that is not paid to the worker is not taken from him but is his quota toward the development of the collectivity. And what is to be produced with this quota are not goods defined as necessarily salable but good that are socially necessary. For this it is essential that a society reconstruct itself in a revolutionary manner if it intends to become a society of workers whose leadership renounces both the tendency to leave everything to chance and the hardening of bureaucracy. (Freire, 1978)

Freire poses the question of what the social needs of the Self and society overall are, and the need for renouncing in a "revolutionary manner." I interpret this as needing to renounce Development and its emergent ideologies of livelihood. Ecopedagogical expansion of this quote, which I would argue would be something that Freire would add himself, is also respecting and loving the rest of Nature toward planetary sustainability and planetary citizenship.

The quote by Shiva discusses how Nature must be understood as supporting "our" livelihoods, and more importantly our survival, with unsustainably as lowering/ceasing livelihood and as possessing the potential for ending our survival.

> Sustainability refers to Nature's and people's sustainability. It involves a recovery of the recognition that Nature supports our lives and livelihoods and is the primary source of sustenance. (Shiva, 1992)

This is specifically of concern when livelihood becomes individualized lifestyles within the realm of neoliberalism and capitalism, with accelerating accumulation, consumption as endless without regard to everyone else's livelihood, coinciding with Dirk Posta's (2006) argument on the private sphere.

Ecopedagogical work centers on *whose* private spheres are affected by the demands of others private spheres, needing pedagogics of collectiveness of private spheres to construct all-inclusive public spheres. Critical GCE work with ecopedagogies has the goal of praxis to determine livelihoods which are globally holistic and planetarily sustainable (Misiaszek, 2018b, 2015). One's lifestyle must be problematized not only within individual spheres but also within societal structures that construct "livelihood" within the conscious and the subconscious. The Freirean scholars below discuss the need for structural change beyond individual actions making up their own lifestyle.

> Structural changes in the society depend on many factors. Human beings are conditioned but not determined. There is a dialectical relationship—unity and the opposition of contraries—between structure and consciousness. Consciousness is not merely the consequence of social, economic, and political structures. Moreover, consciousness alone will not transform the world. It is not enough to be conscious of the world we want, we need to organize, socially and politically, in order to change it. At the PFI [Paulo Freire Institute], we insist on linking the individual dimension of consciousness with the social and collective dimension of organizational work. For instance, it is not enough to change your individual lifestyle to make the world more sustainable. It is necessary to change the system that produces an unsustainable lifestyle; one that produces injustices and inequalities simultaneously. (Antunes et al., 2019)

Ecopedagogical questions emerge from problematizing what individual changes to unsustainable lifestyles (from the individual to societal to global levels) are connected to un/sustainability. This is with the recognition that the term "lifestyle" and the phrase "having a lifestyle" are often seen as hierarchical, as many struggle to meet the basic needs for their survival, with "lifestyle" used here as a non-elitist term.

In addition to commodification, neoliberal globalization has intensified competition, which is related to how one views their livelihood. The problematizing of competition is essential in ecopedagogical deconstruction and reinvention of development and its (un)sustainability. Absolute competition, as ingrained within Western ideologies of neoliberalism, and to a lesser extent within capitalism, constructs zero-sum structured societies in which equality means losing rather than all winning. Once again referencing Freire's first chapter of *Pedagogy of the Oppressed* (2000) in which he problematizes the idea of individual success as how many people are "beneath" an individual, neoliberal competition coincides with such hierarchical "success." Within such a singular goal of success, livelihood becomes hierarchical without solidarity for others and, much less, with the rest of Nature. Also in the first chapter, Freire (2000) discusses the near-impossible task of class suicide, which was previously discussed as an act of giving up all of one's privileges entirely. With this difficulty, the question is what privileges, including aspects of livelihood, are unjust and need to be given up for others not to suffer. Ecopedagogies problem-pose how framings of citizenships can go toward this goal with possibilities, as discussed before, for class suicide partially occurring where they need to. Ecopedagogical work also has to focus on locating the "where" in the previous sentence.

Solidarity has been linked back to aspects of degrowth as how livelihood is framed within neoliberalism in needing to "sacrifice" in such livelihood framings.

> True solidarity demands meaningful sacrifices from those who stand to gain from the existing "sacrifice zones." Degrowth activists would do well to present degrowth in the global North as the only ethical way of responding to and standing in solidarity with environmental justice struggles in the South, when seeking support from EJ [Environmental Justice] movements as Alier (2009); Martinez-Alier (2012) suggests. Presenting degrowth in the North as a form of solidarity with the environmental movements and aspirations for "good life" in the global South opens possibilities for alliance between EJ and degrowth activism on grounds of reciprocal relationality. (Singh, 2019)

Ecopedagogies problem-pose what livelihood is within the constraints of others' social justice and planetary sustainability, but not constrained by oppressions from others. In other words, it is education on development with the aim of increasing the livelihood of everyone, countering oppressions that suppress the livelihood of one's Self as well as all humans together, in balance with the rest of Earth. Within ecopedagogical spaces, I would argue that the deconstruction of what is "sacrificing" in the above quote, including problematizing one's livelihood, must be grounded in needing to increase social and environmental justice for others and planetary unsustainability. I would further argue that toward such ends of ecopedagogical teaching toward deconstructing what is "livelihood" in terms of development and growth, the accompanying deconstruction of what denotes "sacrifice" lends toward eases the impossibly, as Freire (2000, 1998b) argued, of one's actions toward class suicide as the following:

> The essence of my ethico-political choice is my consciously taken option to intervene in the world. It is what Amilcar Cabral called "class suicide" and to what I referred in Pedagogy a/the Oppressed as a resurrection during Easter. In fact, I only diminish the distance between myself and those who are exploited by the injustices imposed upon them when, convinced that a just world is a dream worth striving for, I struggle for a radical change in the way things are rather than simply wait for it to arrive because someone said it will arrive someday. I diminish the distance between myself and the misery of the exploited not with raving, sectarian diatribes, which are not only ineffectual but also make my attempts at communicating with the oppressed even more difficult.

In the same line of arguments, there is the problematizing of what is "class" and, if it is socioeconomic, what framings of economics are we defining/framing it through.

Neoliberalism largely continues due to the teaching for and through neoliberal ideologies and would likely cease if the masses more fully know the truth of the environmental devastation caused as a result of their oppressions—quite literally, a time bomb of ending the world. In the last chapter of his book *Deschooling Society* (1983), Ivan Illich's asks the following question: How do we no longer become a "plaything for scientists, engineers, and planners" (1983) to construct our every want and teach to counter false "common sense" needs? In this sense, the three groups of professionals could be seen as agents for Development, but it is also most often larger systems within which these professionals work. The ecopedagogical focus is on how neoliberal ideologies, as well as other oppressive ideologies such as authoritarianism and untethered capitalism, can lead to perceptions of livelihoods that are unjust and unsustainable.

Notes

1 Lynette Shultz's (2007), three model types as the following:

- Neoliberal Global Citizenship: centers on global understanding for increased economic profit through global expansion and "the role of the individual as an entrepreneur in the private sector is a privileged position."
- Radical Global Citizenship: focuses on defining development as processed towards eliminating global structures that are seen as mostly sustaining the world's inequalities, especially financial ones, that deviates from this definition of development.
- Transformationalist Global Citizenship: views globalizations as both empowering and disempowering with the goal of social justice at the local level by strengthening empowering processes and eliminating oppressive ones—"globalization is viewed as more than a new form of imperialism or just a path to a single global market economy."

2 "Regarding lived experience, it is not possible to distinguish the experience itself from the subject that lives it" (De Sousa Santos, 2018)

3 Identities do not necessarily mean specific experiences, as John Dewey (1963) wrote many years ago that when someone identifies with Antarctica it does not necessarily mean the individual has been to the Artic.

Part Four

Conclusion

Ecopedagogical Possibilities and Challenges

7

Limit Situations of Ecopedagogies

Post-Truthism and the Sustainable Development Goals (SDGs)

This final chapter will be quite short compared to this book's other chapters. With all of the aspects of ecopedagogical work laid out throughout the chapters, I write this chapter not to sum up everything in a step-by-step approach but rather to briefly discuss two current aspects: post-truthism and the UN's Sustainable Development Goal (SDGs). The two are quite different, with the first inherently countering the very foundations of ecopedagogies (as well as critical pedagogies overall) and the second somewhat aligns with ecopedagogies, but not on the same "tracks" as them. Through these chapters I have discussed the various needs of a Kuhnian paradigm shift (i.e., (Kuhn, 1970)) in teaching, reading, and researching environmental violence, which I would argue would be ecopedagogical work—from ecopedagogies, but also ecopedagogical tools with other environmental pedagogies such as EE and ESD. Within the topics of this chapter, I contend that ecopedagogical work is essential to disrupt post-truthism and for the necessary revolutionary transformation to emerge from the SDGs (Misiaszek, 2020, 2019, 2018b, 2015).

7.1 Disrupting post-truthism and shoveling bullshit

In the post-truth era (named *post-truthism* in this book), there are purposefully taught falsities that the very actions that will most oppress various populations will empower them—very much coinciding with Marx's principle of how populations are "best" controlled by hegemony. I have emphasized through some of my most current writings (2020a,b, 2019a), as well as throughout this book, that one of the largest current challenges to ecopedagogies, and also emphasizing the need of ecopedagogies, is the present era of post-truthism.

Before beginning this section, it is important to note that there are many scholars who have argued that post-truthism is not new and should not be seen as such, as it minimizes the critical work on deconstructing histories of false truths. However, there are specifics to intensifying globalizations, especially emergent _T_echnologies (as opposed to _t_echnologies), that have altered approaches and de-/intensified their impact, which I have focused upon.[1] Sheila Jasanoff and Hilton Simmet (2017) have argued the following critique of post-truthism.

> The label "post-truth" signals for many a troubling turn away from principles of enlightened government. The word "post," moreover, implies a past when things were radically different and whose loss should be universally mourned. . . . we argue that this framing of "post-truth" is flawed because it is ahistorical and ignores the co-production of knowledge and norms in political contexts.

The anthropologist Sondra Hale (2018) has made the very important comments below on the validity of blindly accepting post-truthism, especially the methods and practices emergent from such constructs.

> The late historian Hayden White (1973), in the early 1970s, saw history as storytelling, and all stories as fiction. He argued that historical meaning is imposed on historical facts through storytelling.[2] Of course, I could go on and on, but I am simply trying to make the point that the universalism of "truth" and the validity of the concept of "Post-Truth," when brought into focus, are on shaky ground, but arguably not necessarily for the reasons we have been led to believe. . . . Although thinking in terms of post-truth might be a method for interrupting the Modernist agendas, I wonder if these strategies lead us anywhere. And how does any of this relate to gender? Is this post-truth era more difficult for feminists/women teachers/mentors than it is for men?

Although current foci on post-truthism have possibilities of countering oppressions, Hale has expressed that post-truthist concepts are not "new" and has warned that its constructs could further entrench sociohistorical oppressions, with her focus on gender in her keynote, and counter or weaken much needed critical and postmodern work.

Hale (2018) offered various questions in her keynote, including the following one for educators: "Do we have the tools, methods, and critical pedagogical strategies such as Freirean, feminist or decolonialist tools to deal with 'Post-Truth,' as well as the wisdom to deal with the populist atmosphere that pervades?" Hale's sediments and questions, including the one I have highlighted, coincides very closely with the ecopedagogical work that is expressed throughout this book, especially on issues of decoloniality, Development as opposed to development (including coinciding k/Knowledges, s/Sciences, and t/Technologies), epistemological, and ecofeminism. Below are just a few important framings on understanding how falsities that benefit hegemony (e.g., Development) are constructed as truths within ways of knowing, especially within epistemologies of the North.

> Political becomes epistemological when any political alternative to the current state of affairs is credibly framed in the same way as fancy against fact or as falsehood against truth. (de Sousa Santos, 2018)

Boaventura de Sousa Santos' quote equates the politics of knowledges as mendable to become either a perceived "truth" or a "falsity," without the goal of seeking either, but rather the measure of fact/truth to coincide to politically align with previously held ideologies.

Although there are many arguments that post-truthism is not a new "era" but rather has long histories, I want to reemphasize that there are specific aspects of the current era that are unique and will intensify with globalizations from above. Opposingly, globalizations from below provide possibilities to counter such post-truthism's affects as previously discussed, including technologies. Technologies have exponentially increased access to information in various ways, but it also has allowed us to better select information that fits our opinions which, in many ways, have allowed countering knowledges, perspectives, and ways of knowing to be blocked, and diverse thoughts to be increasingly "self-censored."[3] The post-truth era has led to the censorship of facts, truths that conflict with one's ideologies and conceived livelihood(s)/lifestyle(s).

Self-reflectivity of local-to-planetary impacts of actions to sustain one's "livelihood" or self-defined "lifestyle" needs ecopedagogical problematizing of both at their ideological and epistemological roots. In more macro terms of E/SD, such ecopedagogical work problematizes whose livelihood we should sustain, and at what costs. In other words, whose "wants" should be fulfilled over others' wants and, more importantly, needs and survival? One difficulty of problematizing post-truths is the "comfortableness" (i.e., "convenience") of post-truth facts that allow for uncomfortable (i.e., "inconvenient") facts to be ignored. For example, the ability to have endless energy and consumption without needing to worry about global warming, pollution, nonrenewable trash accumulation, fossil fuel supply limits, contribution to social violence, and Nature's destruction, to name a few, is comfortable/convenient to those privileged to use and economically profit from its usage.

So, one key aspect is how we disrupt this comfortability/convenience, whether the "comfortability" is actually true or rather falsely taught, is key to disrupting post-truthism. As with most environmental violence, as discussed throughout this book, the connecting social violence is taught as falsely distanced in the same way that environmental oppressions are distanced as fake, distancing "their" socio-environmental oppressions as devalued from continuing "our" lifestyle/livelihood, and timelines of destruction are distanced for future technological breakthroughs as solving everything. Such distancing too could be argued as comfortable. In the post-truth world, praxis emerging from delegitimization some knowledges and sciences needs rigorous analysis that expands to political opinions, creating facts without Sciences or sciences, or knowledges or Knowledges.

I am going to analyze some of the issues I have discussed in this section inside and outside Plato's cave (2018b). The following passage from Susan Sontag's book *On Photography* (1977) utilized Plato's cave analogy with false truths, which is very relevant in the current post-truth era in various aspects.

> Humankind lingers unregenerately in Plato's cave, still reveling, its age-old habit, in mere images of the truth. (1977)

Within post-truthism, I argue that not only are the shackled prisoners seeing shadows on the cave walls but even the shadows are manipulated realities by shadow makers (i.e., the post-truthists) to coincide with false truths that they want the prisons to "know" as truths. With post-truthism, I would argue that the largest threat is the impending environmental doom without any foundation on the facts that the world can have dialogue for solutions to emerge.

Ecopedagogical tenets emphasize the need for rigorous praxis for effective solutions to emerge, but post-truthism systematically suppresses such decision-making leading toward action and even allows shallow, noncritical problem-solving of environmental issues. However, there is also the argument that post-truthism with globalizations have also allowed us to be more comfortable in our own opinions by allowing us to cocoon ourselves in with like-minded people through selective website searches and social media. Such comfort comes from avoiding conflict, which also prevents authentic ecopedagogical dialogue. When lying becomes a strength and s/Scientific truths are seen as a weakness, the world (anthropocentric sphere) eliminates possibilities of conflict dialogic within learning spaces. This is because dialogue no longer has the groundings of some commonly held truths, and discussions are aligned with speaking to self-ideologies rather than listening (Misiaszek, 2020). Tracy Bowell explains this through Plato's cave allegory, with not wanting to leave our constructed caves with like-minded others.

> These echo chambers shelter us from socio-cultural and political difference; their digital (dis)locatedness shields us from an encounter with the distant other, who is demonised in the absence of any challenge to our prejudices and fears that could be gained from the experience of an actual, embodied, encounter. As Plato reminded us, the illusions of the cave offer more comfort than the confrontations of reality. Lorraine Code's (1984) Aristotelian characterisation of the intellectually virtuous person shows the epistemic flaws inherent in our tendency to remain within the shelter of the cave. (Peters et al., 2018) (2018)

Ecopedagogical work cannot be done without leaving the/our cave, but rather requires challenging our knowledges and ways of knowing to better understand the world-Earth. In the same sense, Paulo Freire (2000) discussed the fear of freedom as the unknown of how the Self is understood by the Self and others, countering normative narratives of _D_evelopment, as an example.

_D_evelopment is reassuring to many owing to its familiarity, even though it oppresses the vast majority who are reassured by it. Post-truthism might seem like it is countering oppressive, dominant structures and systems within and between societies, but those structures and systems are often the puppeteers of feeding false truths (i.e., the shadow makers). This could be exemplified by my arguments of the "comfortableness" and "convenience" of livelihood/lifestyle of _D_evelopment previously discussed only for a few and many others to fatalistically aspire to as it is not "for them." There are many arguments within ecopedagogical work countering post-truthism, but the foundational one is that socio-environmental praxis—or critical, transformative praxis overall—cannot emerge from teaching, reading, or research based on false truths that are portrayed as truths. An essential aspect is how to ecopedagogically teach and read to determine "truths" as false or true. As stated throughout the book, ecopedagogical work reads and rereads socio-environmental and planetary "truths," but rather than the goal of following an opinion and/or an ideology, the goal is to identify true truths—a key distinction between post-truthism and the subjectivity of critical and postmodern praxis (Misiaszek, 2020).

Post-truthism has long-term negative effects that question what we know as "true," in that it disregards _s_/_S_ciences toward the truth that select facts without looking at the larger picture. For example, with more than 97 percent of the US climatologists agreeing that humans are altering the climate (Cook et al., 2016, 2013; Anderegg et al., 2010; Doran and Zimmerman, 2009; Oreskes, 2004), post-truthism questions why not 100 percent, without the understanding that much of sciences is not 100 percent (e.g., 100 percent of doctors still do not believe cigarette smoking is necessarily unhealthy, and in recent studies 11 percent of people think vaping is harmless,[4] and around 1.28 to 2 percent of the people believe Earth is flat[5]). In my courses, I often ask my students, "How many of you believe that smoking cigarettes does not lead to lung problems?" Although some of my students might smoke, it is, without question, difficult to state that it is "healthy." The following quote expresses how humans can become increasingly literate in reading post-truths as falsities, despite our collective fears.

> The human animal may be wired to give some credence to our superstitions and fears, but this does not mean that we cannot train ourselves to embrace better standards of evidence. (McIntyre, 2018)

With climate change, it is questioning not only the science but that the science was systematically manufactured by those who want to hurt the United States, as Trump suggested that China made up climate warming to hurt the US economy. Within ecopedagogical literacy, the question is, how can we teach to read falsities, including those that lead to socio-environmental violence, including the politics of the world with the objective laws of Nature?

Flawed arguments that post-truthism is just utilizing critical and postmodern arguments that all knowledges, sciences, and research are political disregards differing goals of truth or "convenient" falsities. Such shallow readings of critical/postmodern work are devoid of efforts to seek out the politics toward "truer" truths; they rather mystify truths through the use of politics for propping up false truths. For example, why did Trump bring up China while more than 97 percent of the US climatologists would have, by a vast majority, more interest in the well-being of the United States, rather than China? His tweet helps intensify some of the US fears of China's growth as a global economic power, and links climate change to this fear rather than having anything to do with any true facts. With the argument that this tweet from Trump, and way too many other tweets from him, are not a result of ignorance but rather aim at systematic manipulation of connecting unrelated, manufactured fears in halting climate warming as negatively affecting the US economy. The tweets are all out of any realm of seeking what is true and seeking possible solutions to stop the ultimate disaster that climate warming will have on the world and Earth, including the US economy.

If ecopedagogical research goals are to more methodologically answer socio-environmental curiosities with truths and educate people on these truths, ecopedagogical researchers have three roles to play: first is for their work to better understand the truths and their contextualities; second is to use their research toward helping those researched; and third, which has been increasingly important, is to teach to counter false truths that portray their research outcomes as falsities and lies. It is this third one which might be the most complicated aspect in an increasingly globalized post-truth world, and could somewhat coincide with Harry Frankfurt's (2005) framing of "bullshit."

Princeton Emeritus Professor of Philosophy Harry G. Frankfurt's book *On Bullshit* (2005) begins by stating that there is so much bullshit in the world which everyone knows and accepts, to such a large degree that it's largely unquestioned and has little to no research or theorizing done upon it. Will post-truthism largely go the route of bullshitting, as a more extreme version, in that telling the "truth" will no longer be expected as normal and lying will become the norm? Frankfurt (2005) discussed throughout his book the differences between bullshitting and lying, with the following combined quote as a good summary.

> Bullshit need not be false, it differs from lies in its misrepresentational intent. The bullshitter may not deceive us, or even intend to do so, either about the facts or about what he takes the facts to be. What he does necessarily attempt to deceive us about is his enterprise. His only indispensably distinctive characteristic is that in a certain way he misrepresents what he is up to. . . . [it is] impossible for someone to lie unless he thinks he knows the truth. Producing bullshit requires no such conviction. A person who lies is thereby responding to the truth, and he is to that extent respectful of it.

The intention of lying is key to post-truthism which differs from bullshitting, as Frankfurt describes it above—with the intent and knowledge of lying. There are a few aspects of post-truthism to consider here, first systematically, it is lying within this definition; however, those who spread post-truths (e.g., through social media) are not necessarily knowing the truth, and thus they would not be considered liars within Frankfurt's framing of bullshit. Rather, it is those who systematically spread false truths, who fully know that the truths they are spreading are false, are in fact lying and utilizing others whom they indoctrinate with the falsities as truths, and spread such knowingly false understandings. This could very closely coincide with Marx's argument on how best to control populations that was discussed at the beginning of this section.

Taking, for example, again Trump's tweet on the Chinese government creating climate change, does he believe this to be true? I would say no, although maybe something presently is unknown that will come out later, such as a mental defect or disease. Why does he tell this lie then? One reason would be to flame fears of environmental regulation, economic downturns, and China's rising economic influence globally, as discussed previously as feeding on non-connective false fears. Second, acknowledging climate change would counter his campaign promises to help disenfranchised populations in the fossil fuel industries, especially coal miners, to build up their industries again (an impossible and socio-environmentally frightful goal). Third, within neoliberal framings, in which everything comes down to economics and hegemony, environmental concerns are not neoliberal concerns of current powerful interests as they diminish the fact that they are economic concerns in the not-so-long-term future, if not "recognized"/"felt" already. Fourth, to spread extreme nationalism and xenophobia of the non-Westerner, and more specifically the non-"American,"[6] which will be discussed more at length later in this section with citizenship. Overall, climate-change facts counter his goals, but the false tweets help to further ignite and connect the two following aspects that he wants to instill in people's minds: fear of China and fear that climate change is bringing about changes that will hurt *D*evelopment (veiled as *d*evelopment) and "your" livelihood/lifestyle.

Although a vast majority does not believe what he has tweeted, it further entrenches unconscious aspects of the "underlying belly" of China's deceit (i.e., fear of China), that climate change is not true, and diverts people's attention to all else that is going on by the sheer idiocy and vast number of tweets like this he and his administration send out. In other words, if people focus on these tweets, many of the more substantial actions that sustain/intensify climate change goes on with less notice, resulting in less protest and resistance against such environmental violence. Trump would later say that he was joking—as a form of bullshitting rather than lying. This is a common technique used by Trump to further him from needed scrutiny, but the false knowledge is already out there, within people's subconscious, and his admission is much, much less prominent in the media than his initial tweets. This follows Frankfurt's (2005) statement that this "presumes not only that there is an important difference between lying and bullshitting, but that the latter is preferable to the former." In his book, Frankfurt (2005) provides an in-depth discussion on the sociohistorical factors defining bullshitting, with one that is clearly obvious by observing Trump, among others, in that people who are supposed to be knowledgeable in a certain area will most likely bullshit the most when asked about situations unknown to them. Trump relying on bullshitting could be related to what Frankfurt (2005) described as the need to bullshit to "being true to his own Nature . . . as though he decides that since it makes no sense to try to be true to the facts, he must therefore try instead to be true to himself" (Frankfurt, 2005).

Trump's tweeting, I would argue, is one of the strongest examples of *T*echnological usage of social media. It highlights answering the question whether post-truthism is an era different from other points of history in that there are more effective tools for false-truth telling, such as social media. In *On Bullshit* (2005) Frankfurt points out that it is very likely that it has not increased but rather communications have increased, including the access to communications witnessed back in 1986 (when initially written), and such communications and access to falsities/truths have exponentially increased with the internet. The following quote also discussed how increased bullshit might also be a product of anti-realistic deconstruction and reconstruction of "facts."

> Contemporary proliferation of bullshit also has deeper sources, in various forms of skepticism which deny that we can have any reliable access to an objective reality, and which therefore reject the possibility of knowing how things truly are. These "antirealist" doctrines undermine confidence in the value of disinterested efforts to determine what is true and what is false, and even in the intelligibility of the notion of objective inquiry (Frankfurt, 2005)

The above quote could return to the notion that aspects of postmodernism, or to a lesser extent critical theorizing of the politics of sciences and knowledges, have intensified bullshitting and, therefore, lead to the current post-truth era.

Returning to the topics of extreme nationalism and xenophobia of noncitizenships framings, post-truthism has intensified anti-globalization based on hate and fear rather than globalization toward ending oppressions emergent from global hegemony. There are endless cases of this but a key one is support for Donald Trump getting elected as the US president through fears and hatred for the othered, such as "immigrants" as the noncitizen to fear and hate, rather than oppressions from transnational corporations (TNC)—in which his work and, many would argue, his only interests are located. Robert J. Tierney has discussed the need for global meaning making to counter extreme nationalism and associated unwarranted forms of privilege which inherently counters globally inclusive socio-environmental justice and planetary sustainability.

> Critical global meaning making is tied to challenging isolationism, protectionism, nationalism, and forms of exceptionalism and unwarranted privilege. Rather than perpetuating the reader's compliance with a global inequitable status quo, global meaning making interrogates whose interests are served by global developments and affordances . . . challenging isolationism, protectionism,

> nationalism, and forms of exceptionalism and unwarranted privilege. (Tierney, 2018b)

Tierney further discussed the need for global meaning making to deepen the understandings of "intentions and possible effects" as essential, especially within post-truthism, with the following quote.

> [Global meaning making] requires being vigilant about what is presented by whom to whom, why, when, and where—cognizant of the intentions and possible effects . . . global meaning making purports to confront the regressive tendencies of the current "post-truth" world, especially the insularity of approaches to ideas that advance the suppression of environmental and social developments and the walling off truth with unfettered hyperboles, misrepresentation, oppressive personifications, and claiming false equivalencies. (Tierney, 2018b)

Within the focus of unsustainable, environmentally violent actions, the deepened understandings coincide with deconstructing and denouncing post-truthism's resulting in "unfettered hyperboles, misrepresentation, oppressive personifications, and claiming false equivalencies."

Post-truth*ism* adds a difficult dimension of ecopedagogical work to disrupt such resultants emergent from false truths grounded on opinions and Development ideologies. With the questioning of the critical and postmodern theories as inciting post-truthism, critical/ecopedagogical reading must include if one's "own ideological litmus test to an area of inquiry (even though they may deny that this is what they are doing) under the guise of 'openness' and 'fairness'" (McIntyre, 2018) is being applied rather than seeking truth through destructing ideological framings. An essential problem-posing question for those who deny global warming would be "Am I denying climate change because I/others benefit from it not being true?" Other self-reflection includes problematizing if "fairness" and "openness" to question if climate warming as real emerges from self-ideological alignment or from seeking the "truth," including through problematizing ideologies, including the ones of one's self? Teaching environmental issues becomes increasingly complicated with globalized mobility/transferences between local, national, and global knowledge systems, religious/spiritual beliefs, and Nature's laws. With an increased emphasis on "scientific" measuring, especially resulting from increased positivism within neoliberal globalization (Stromquist, 2002b; Torres, 2009), another question of how we construct "being scientific" becomes increasingly important.

The key factor of all teaching and research should be going toward "truths," with the critical aspect of there being few fully objective truths that are not subjective or contextual as being "true." As discussed throughout this book, the laws of Nature within the non-anthropocentric sphere would coincide with such absolute truths, with most other aspects of the world being subjective. This second part is where many scholars have given the aspects on how the subjectivity of critical theories and the more intense subjectivity of postmodernism has led toward intensified post-truthism. However, many of the same scholars argue that most post-truthers are not using academic scholarship, such as through Foucault's arguments, to argue the needs for their post-truthist work (without naming it "post-truth" work).

As post-truthism's goal is not toward "truth" but rather toward false truths for political gain, post-truthism, I have argued (Misiaszek, 2020), is the key aspect of being post-truth*ist* within environmental pedagogies and pedagogies on the environment. In other words, the recognition of subjectivity and limitations of knowledge (including "knowing" Nature, as discussed previously) is within ecopedagogies, but it is the intention of teaching and research that is essential (i.e., toward or away from truth seeking). Goals of falseness are why the counterargument below that McIntyre (2018) poses, as countering his own views on post-truthism, are not valid is just having another way of thinking, another epistemology, or "alternative facts" as coined by Trump's administration.

> "Post-truth" is irreducibly normative. It is an expression of concern by those who care about the concept of truth and feel that it is under attack. But what about those who feel that they are merely trying to tell the "other side of the story" on controversial topics? That there really is a case to be made for alternative facts? The idea of a single objective truth has never been free from controversy. Is admitting this necessarily conservative? Or liberal?

Alternative facts are emergent on their popularity even though "virtually all credible sources would dispute" them, in that in their "purest form, post-truth is when one thinks that the crowd's reaction actually does change the facts about a lie" (McIntyre, 2018). Critical pedagogies, including ecopedagogies, must question truths within K̲nowledges, S̲ciences, T̲echnologies, D̲evelopment, and within epistemologies of the North; however, the goal is for truths that deconstruct the falsities, and accompanying falsities, that sustain socio-environmental injustices and unsustainability. There are numerous conundrums as discussed previously in which the objective laws of Nature might not coincide with aspects of social justice and livelihoods, but these are complexities to deal with. Within post-truthism, we just falsify and adjust the objective and *un*adjustable laws of Nature—which is impossible, leading to eventual ruin and further de-democratization of our populations.

Pedagogies on the environment increasingly teach an inherent reverse relationship between livelihood and progress with environmental well-being (Misiaszek, 2011), directly linking to unsustainable Development ideologies (Misiaszek, 2020, Misiaszek and Torres, 2019; Misiaszek, 2020; 2018b). Analyzing livelihood is essential in constructing development by deconstructing neoliberalism's and post-truthism's singular centralization of one's Self's Development (Misiaszek, 2020). Neoliberal framed education devalues environmental well-being as negatively affecting one's own private and public spheres, with the individual sphere as solely centered in neoliberal ideologies (Misiaszek, 2016a; Postma, 2006). Among the other aspects of post-truthism discussed that coincide with neoliberal*ized* livelihood, the education of false truths diverts the "true" causes of one's decline of livelihood to those that have opposing ideologies, "others" that are "distanced" from the Self, falsifying knowledges/sciences that prove socio-environmental connections, and "illegitimate" framings of humans as effecting Earth and/or planetary constructs, among other false reasoning frameworks (Misiaszek, 2020).

Ecopedagogical teaching must be through problem-posing of false reverse relationships by asking how we can define livelihood, progress, and development that stresses this reverse relationship and what the politics behind such teachings are. Opposingly, post-truthism has one ideological endpoint without concern for verified facts and inherently countering such dialogue by ignoring all knowledges and epistemological framings that oppose current opinions/ideologies to discuss environmental violence through lenses of falsities. This includes viewing one's own livelihood without considering others' livelihoods outside of their own self-defined population(s). In short, post-truth populism communicates false goals for empowerment against hegemony, as post-truthism structurally sustains and intensifies that same hegemony.

Those who believe false truths without critical reflection are within the political processes for oppressing those which the false truths aim to attack, as well as to oppress the "believers' themselves" The believers are fundamentally de-democratizing themselves, as their decision-making and subsequent actions are grounded in falsities that often socio-environmentally oppress them.

With all this being stated, increased k/Knowledges and overall education do not necessarily lead to less believing in false truths from the mechanisms of post-truthism's, but rather how teaching is done. There are many obstacles, including one of the most important that increased knowledge, even when is true, does not necessarily lead to someone knowing what is factual as compared to what are false truths (Frankfurt, 2005). This becomes a difficult obstacle within banking education models with the belief that sheer increases of knowledges will lead to better understandings. But, also in non-banking models, teaching the tools to understand and use the knowledges for problem-solving does not necessarily lead to countering false truths. Further complications include the arguments that increased education does not mean that a person becomes less susceptible to post-truths, but rather often it could mean the opposite (Frankfurt, 2005; McIntyre, 2018). In this respect, Lee McIntyre (2018) argued that we "feel better for us to think that we are smart, well-informed, capable people than that we are not" (McIntyre, 2018). Such reflections of one's Self becomes problematic when we think of ourselves too smart to be duped, countering our tribalism of being one "in the know" within a relatively small collective. Some think falling for post-truthism is due to lack of education; however, there are counterarguments on if education is a direct factor in lessening beliefs in false truths because the more educated someone is the less they will question their own previous knowledges, political beliefs, and identities (Frankfurt, 2005; McIntyre, 2018). I often self-reflect upon myself to my students that, according to this argument, the most prone to post-truthism is myself in the former and "us" in the latter situation. I, of course, don't think "I" am, but stating this might prove the point further.

Such alignment with the Self's beliefs is beyond a binary of believing or not believing "facts," but also the indirect agreements toward aligning with what we "know" as "true"—as to "preserve our sense of self-value" (McIntyre, 2018) by seeking confirmation and dismissing truths that counter our beliefs (Frankfurt, 2005; McIntyre, 2018). Rationality of the Self and corresponding shallow self-reflectivity most often leads to confirmation rather than critical problem-posing (e.g., noncritical literacies and non-ecopedagogical literacy).

As discussed earlier in this section, one of the main points of contention of critical and postmodern theories is the level of subjectivity, with difficult questions with respect to whether oppressions and suffering are always "real" or not (a very important, but not the only important, question to problem-pose here). For example, returning to Hayden Whites' (1987, 1973) argument that "all history is fiction" helps to deconstruct perspectives and systematic ideologies of histories and linguistics; however, what happens when a resulting argument is that the Holocaust was fiction. Such questioning of suffering is both inside and outside academic arguments, with experiential, bottom-up constructed knowledges and sciences as essential. Within ecopedagogical work, such extreme subjectivity could question environmental disasters, such as the Union Carbide industrial disaster in Bhopal, India, which killed 3,800 to 16,000 people with over 600,000 very negatively affected (Taylor, 2014). It is also important to note that not only these large catastrophic events but all the environmental violence that does not grab the headlines, building up toward world-Earth destruction, are also equally devastating. The significance of this can be related to the frog analogy that frogs will quickly jump out of water that is already boiling but will remain in water if the water is slowly heated to the boiling temperature, implying that we, in becoming knowledgeable, should jump out of the soon-to-be boiling water, like the frog, rather than waiting for the world's destruction.

7.2 Interrogating the Sustainable Development Goals (SDGs)

This is not going to be a very substantial section on ES/D seeing that this entire book focused on this topic throughout and any simplification here would narrow down the complex arguments throughout this book. The United Nations Sustainable Development Goals (SDGs), as well as the Millennium Development Goals (MDGs) and whatever follows the SDGs, has the likelihood of not being fully achieved for all the world-Earth.

One of the key aspects of ecopedagogical reading, teaching, and researching global initiatives (e.g., SDGs) is determining the underlying politics of what environmental actions are promoted and, I would state more importantly, what actions are systematically discouraged, and why. In this section I will briefly discuss the SDGs as, in part, having environmental sustainability goals but prioritizing the aspect of sustaining "development" more within dominant systems globally—sometimes Development but not always. In an article on ecopedagogy and stoicism that I coauthored, we wrote the following on the SDGs and sustainable development in general.

> The distinction between virtues (which constitute absolute "good") and values is apparent in the oxymoronic promotion of the UN's SDGs, many of which hinge upon national Gross Domestic Product (GDP) and the notion of "sustained growth." This can be identified by posing the following questions related to the discourse of sustainability: "Whose interests does a sustainable development based on economic growth really promote? And, at whose or what expense?" A traditional or indigenous community may find that in the name of "sustainability," their natural environment and social fabric have become devalued or demoted as secondary to "economic wellbeing" (Raworth, 2017). This is a complex problem frequently discussed by Freirean ecopedagogues such as Misiaszek (2018b),

> Misiaszek and Misiaszek (2016) and Gadotti and Torres (2009). The fact that such a contradiction exists in the UN's SDGs demonstrates just how entrenched neoliberal ideology is in modern society and how engrossed societies are by growth and capital accumulation (Carmona et al., 2017). Such value judgements and actions are likely to be antithetical to sustainable development and actually undermine its progress. This is why it is important to evaluate development policies through a critical lens. (Whiting et al., 2018)

One of the key questions of development and sustainability is how these terms are taught—through what epistemology(ies), theoretical lenses, within what framings of values, moralities and ethics, and through whose histories. Does the global initiative of the UN's SDGs allow for such deepened ecopedagogical work for them to be meaningful and successful in the widened global sphere? This is a key question.

The SDGs and UNESCO's Decade of Education for Sustainable Development (DESD) have shifted development outside of economics to one that includes environmentalism. David Post and Yi Meng (2018) have argued this, with the concluding arguments from David Orr (2009).

> Through the SDGs, the United Nations is trying to integrate—perhaps even reconcile—disparate development targets into a coherent framework for future decision-making. Previously education was considered a route to economic productivity; now it is seen as a path for sustainability. But it is necessary to ask how likely it is that universalizing secondary education will promote development that is sustainable, as opposed to promoting other well-known outcomes of expanded schooling, especially increased income and consumption. The problem, as critics note, is that existing schools can only be part of the solution if they teach the values and skills needed for environmentalism. For example, David Orr (2009) has argued that "... Education has long been a part of the problem, turning out graduates who were clueless about the way the world works as a physical system or why that knowledge was important to their lives and careers, while at the same time promoting knowledge of the sort that has fueled the destruction of ecologies and undermined human prospects."

My overall problem-posing is, if this "shift" is the Kuhnian paradigm shift that is needed for all-inclusive-worldwide socio-environmental justice and planetary sustainability.

As one of the reviewers of the UNESCO publication *Education for Sustainable Development Goals: Learning Objectives* (2017), I made extensive comments that were grounded in critiquing the language of SDGs and the learning objectives in defining "development" and "sustainability," with the economic goals of capitalism, and all too often aligning with neoliberalism, within the wording. Another key aspect that I commented upon was what was not in the goals or the learning objectives, especially in terms of being careful not to disrupt the current economic and power dynamics. The Goals' and the learning objectives' writing lacked the critical emphasis that, for success by 2030, all the goals everywhere must be met, because if they are not achieved anywhere, the Goals will not be achieved overall. It is their success in the world-Earth holistically that is essential for planetary sustainability and achieving socio-environmental justice within the world. Overall the official wording of more "official" SDG literature is progress but it is far from the paradigm shift needed.

Throughout this book, as well as throughout all my ecopedagogical work, I have argued for the need for ecopedagogically problem-posing the framings of development and sustainability (both together and separately) for deepened and widened understandings. Problem-posing for what and whose progress we commit acts of environmental violence in the name of development to critically unveil the hidden politics of socio-environmental injustices and planetary sustainability is essential. Does the SDGs do this or, a better question would be, how can we ecopedagogically reinvent the SDGs to be S*d*Gs rather than S*D*Gs?

I will end this book by giving a few, key aspects of teaching, reading, and research for ecopedagogical praxis toward ending unsustainable environmental violence causing environmental migration, social injustices, and dominance over the non-anthropocentric sphere. Throughout this book I have delved into various aspects of ecopedagogies, some in depth and others not as much, but this book only represents the foundational aspects of the field, with much more rigorous theorizing, research needing to be done. Taking the example of environmental migration, there are endless aspects of education that need to be expanded upon. But the overall goal of ecopedagogical work is to deepen and widen understandings of environmental violence for praxis.

There are various global sustainable development concerns that make the success of SDGs highly important for our world and Earth, as we know it, to continue.[7] Returning back to the example of global warming, the issues surrounding environmental migration is shocking, including forced migration before the SDGs are to be "accomplished."[8] There are numerous figures given on environmental migrants who will be displaced in the next few decades, all of them are disheartening. Various scholars (Lister, 2014; Piguet, 2013; Findlay, 2011; De Haas, 2010) have cited a wide range of predicted estimates of environmental migrates/refugees, with the numbers given being in the tens to hundreds of millions by 2050 and eclipsing all other force migration categories. According to the United Nations' International office of Migration (IOM) there will be between 25 million to 1 billion environmental migrants by 2050, which has the uncertainties of a fortyfold difference between the two numbers (International Organization for Migration (IOM), 2014). The IOM (2014) gives three factors in its final statistics for 2050:

> a) which climate change scenarios will be borne out; b) what adaptation actions are undertaken; and c) the evolution of various socioeconomic, political and demographic factors influencing the decision to migrate, such as economic growth and development, population growth and governance.

The second one asking "what adaptation actions are undertaken" is the goal of ecopedagogies, which delve into the deepened and widened perspectives of the third one, with the specific aspects of economic growth, development, and governance.

Within environmental migration, the contextuality of migration, including environmental migration, which "is not an exogenous variable, but an integral part of wider social and development processes, the development impacts of migration are also fundamentally heterogeneous" (De Haas, 2010). This importance is exemplified in the debates of the label environmental "migrant," reflecting "adaptiveness" to environmental violence rather than the "implicit hopelessness and defenselessness of the term [*refugee*]" (Baldwin, 2017; Adamo, 2010). McGregor's (1993) argument that the term "environmental refugee" is "a false separation between overlapping and interrelated categories" (McNamara, 2007) highlights the inseparable connections between social and environmental issues.

Problematizing the foundations of citizenships is essential to analyze environmental migration: How do we teach and understand the "environmental migrant" as a fellow citizen or a noncitizen, and feel solidarity *with* them or not? Also, what citizenships are these environmental migrants of, as the world has forced them into migration due to unsustainability in the name of Development not for them but for others? What SDGs, including the connections between them, will end this? Do the SDGs allow for radical, revolutionary ecopedagogical teaching and emergent praxis that counters Development goals of those states that are most powerful in the UN? *The Guardian* had an article by Liz Ford (2015) back in 2015 that asked and answered the following question on "happiness" of such states with regard to the SDGs.

> Are governments happy about the proposed 17 goals? The majority seem to be, but a handful of member states, including the UK and Japan, aren't so keen. Some countries feel that 17 goals are too unwieldy to implement or sell to

> the public and would like a narrower brief. Or so they say. Some believe the underlying reason is to get rid of some of the more uncomfortable goals, such as those relating to the environment. Britain's prime minister, David Cameron, has publicly said he wants 12 goals at the most, preferably 10. It's not clear, though, which goals the UK government would like taken out if they had the choice.

I would be very confident in stating that Trump now would have a stronger response than Cameron. With such an assertion, what is the "teeth" of the SDGs, including the education goals of SDG #4 ("Ensure inclusive and equitable quality education and promote lifelong learning opportunities for all"),[9] for radical transformation needed?

An essential educational aspect of, and emergent from, the SDGs (#4 in particular) is how we de-distance one another of the world and "us" with the rest of Earth—thus, increased and thickened world-Earth solidarity. Throughout this book, I have argued that ecopedagogy is an essential element of citizenship education and vice versa, as well as how we determine our actions for "development." Teaching to both widen and localize (or deepen) who/what we consider fellow citizens, helps to increase the connections we have with one another, as "fellow citizens," when socio-environmental oppressions endured by those at large distances. Ecopedagogical practices and research problematize globalization<u>s</u> as it affects our constructs of citizenship (plural to indicate the contested terrain of globalization processes from above and from below) on its socio-environmental effects, specifically on the local level of societies and populations. I have argued that citizenship education must focus on both social and environmental justice, due to the fact that neither will progress if the other is absent, seeing that they are dependent on one another. I have provided some key needs to problematize socio-environmental issues within and between multiple citizenship spheres in teaching and research, as well as initially discuss some of the (dis)connections between ecopedagogy and Global Citizenship Education (GCE) from additional research I have conducted (initially described in Chapter 1). This book also discussed various aspects and needs of planetary citizenship (beyond human-centered concerns) by teaching and reading how negative environmental issues of one community affects us all globally and holistically as we are all part of the planet along with the rest of Nature, no matter the distance (such as geographically, culturally, traditionally, linguistically, or form of matter). How such connections between constructs of citizenships in how "development" is framed will be critically analyzed, both as current connections and how such problematizing must be centralized in ecopedagogical teaching and research.

Notes

1 Systematic post-truthism is "produced" in various ways, including having a "form of ideological supremacy" (Mcintyre, 2018) and tribalism for those who feel disenfranchised by current societal structures, selecting specific ancillary "facts" to focus upon in constructing false truths, governance though falsities without repercussion, and the internet's ability for users to select knowledges that "fit their ideologies" with social media with repetitive feedings for individualistic selection of "knowledges," leading to *source amnesia* (ibid.).

2 White (1973).

3 "Self-censored" is in quotations to indicate that much of censorship online that seems to be from the individual is actually systematically controlled, such as systematically selected Facebook's links to keep users, the Self, scrolling down regardless of if the link are true or not.

4 See http://newsroom.ucla.edu/stories/how-safe-is-vaping

5 See https://blogs.scientificamerican.com/observations/yes-flat-earthers-really-do-exist/

6 I write "America" here with recognition of the linguistical problems of there being multiple Americas (e.g., South, Central, and North America) with the hegemonic wording of "America" associated with the United States and North America, that Lauren Ila Misiaszek has discussed previously in lectures and conference halls.

7 Wording here indicates that the rest of Earth will continue with or without the world (i.e., "us" as human populations).

8 "Accomplished" here is in quotations to signify the unlikelihood of the Goals being fully reached in their entirety, but rather seen as utopic goals.

9 See http://sustainabledevelopment.un.org/sdg4

References

Abdi, A. A. 2008. De-subjecting subject populations: Historico-actual problems and educational possibilities. *In*: Abdi, A. A. & Shultz, L. (eds.) *Educating for human rights and global citizenship*. Albany, State University of New York Press.

Abdi, A. A. 2015. Decolonizing global citizenship education: Critical reflections on the epistemic intersections of location, knowledge, and learning. *In*: Abdi, A. A., Shultz, L. & Pillay, T. (eds.) *Decolonizing global citizenship education*. Rotterdam, Sense Publishers.

Abdi, A. A. & Shultz, L. 2008. Educating for human rights and global citizenship: An introduction. *In*: Abdi, A. A. & Shultz, L. (eds.) *Educating for human rights and global citizenship*. Albany, State University of New York Press.

Adamo, S. B. 2010. Environmental migration and cities in the context of global environmental change. *Current Opinion in Environmental Sustainability*, 2, 161–5.

Adams, F. & Horton, M. 1975. *Unearthing seeds of fire: The idea of Highlander*, Winston-Salem, J. F. Blair.

Adorno, T. W. 1998. Education after Auschwitz. *In*: Adorno, T. W. (ed.) *Critical models: Interventions and catchwords*. New York, Columbia University Press.

Alaimo, S. 2008. Ecofeminism without nature? Questioning the relation between feminism and environmentalism. *International Feminist Journal of Politics*, 10, 299–304.

Alier, J. M. 2009. Socially sustainable economic de-growth. *Development and Change*, 40, 1099–119.

Anderegg, W. R. L., Prall, J. W., Harold, J. & Schneider, S. H. 2010. Expert credibility in climate change. *Proceedings of the National Academy of Sciences*, 107, 12107–9.

Andreotti, V. & Souza, L. M. T. M. D. 2012. *Postcolonial perspectives on global citizenship education*, New York, Routledge.

Antunes, Â. B., Pini, F., Padilha, P. R. & Couto, S. 2019. Fertilizing the unusual (The Praxis of a Connective Organization). *In*: Torres, C. A. (ed.) *Wiley handbook of Paulo Freire*. New Jersey, Wiley-Blackwell.

Appadurai, A. 1996. *Modernity at large: Cultural dimensions of globalization*, Minneapolis, University of Minnesota Press.

Appadurai, A. 2001. Grassroots globalization and the research imagination. *In*: Appadurai, A. (ed.) *Globalization*. Durham, NC, Duke University Press.

Apple, M. W. 2000. *Official knowledge: Democratic education in a conservative age*, New York, Routledge.

Apple, M. W. 2004. *Ideology and curriculum*, New York, Routledge.

Apple, M. W. 2006. *Educating the "right" way: Markets, standards, God, and inequality*, New York, Routledge.

Apple, M. W. 2014. *Official knowledge: Democratic education in a conservative age*, New York, Routledge.

Apple, M. W. & Au, W. 2009. Politics, theory, and reality in critical pedagogy. *In*: Cowen, R. & Kazamias, A. M. (eds.) *International handbook of comparative education* (Vol. 22, pp. 991–1007). Dordrecht, Springer Science + Media Studies.

Apple, M. W., Au, W. & Gandin, L. S. A. 2009. Mapping critical education. *In*: Apple, M. W., Au, W. & Gandin, L. S. A. (eds.) *The Routledge international handbook of critical education*. New York, NY, Routledge.

Arnove, R. F. 1980. Comparative education and World-Systems analysis. *Comparative Education Review*, 24, 48–62.

Arnove, R. F. 2007. Introduction: Reframing comparative education: The dialectic of the global and the local. *In*: Arnove, R. F. & Torres, C. A. (eds.) *Comparative education: The dialectic of the global and the local*. 3rd ed. Lanham, MD, Rowman & Littlefield.

Arnove, R. F., Torres, C. A. & Franz, S. 2013. *Comparative education: The dialectic of the global and the local*, 4th ed. Lanham, MD, Rowman & Littlefield Publishers.

Aronowitz, S. 1998. Introduction: Pedagogy of freedom: Ethics, democracy, and civic courage. *In*: Freire, P. (ed.). Lanham, MD, Rowman & Littlefield.

Asara, V., Otero, I., Demaria, F. & Corbera, E. 2015. Socially sustainable degrowth as a social–ecological transformation: Repoliticizing sustainability. *Sustainability Science*, 10, 375–84.

Assié-Lumumba, N. D. T. 2008. Higher knowledge and global good: Reconceptualizing and envisioning higher education in Africa for shared and enhanced humanity. *2008 General Assembly of the World Academy of Art and Science*. Hyderabad, India.

Au, W. 2009. Fighting with the text: Contextualizing and recontextualizing Freire's critical pedagogy. *In*: Apple, M. W., Au, W. & Gandin, L. S. A. (eds.) *The Routledge international handbook of critical education*. New York, NY, Routledge.

Au, W. W. & Apple, M. W. 2007. Reviewing policy: Freire, critical education, and the environmental crisis. *Educational Policy*, 21, 457–70.

Bajaj, M. 2008. *Encyclopedia of peace education*, Charlotte, NC, Information Age Pub.

Baldwin, A. 2017. Resilience and race, or climate change and the uninsurable migrant: Towards an anthroporacial reading of "race". *Resilience*, 5, 129–43.

Beck, U. 1992. *Risk society: Towards a new modernity*, London, Sage Publications.

Becker, C., Faber, M., Hertel, K. & Manstetten, R. 2005. Malthus vs. Wordsworth: Perspectives on humankind, nature and economy. A contribution to the history and the foundations of ecological economics. *Ecological Economics*, 53, 299–310.

Bishop, A. 2002. *Becoming an ally: Breaking the cycle of oppression in people*, Halifax, N.S., Fernwood Pub.

Bohorquez, I. 2008. Untested feasibility in Paulo Freire: Behind the profile of a dream. *In*: Torres, C. A. & Noguera, P. (eds.) *Social justice education for teachers: Paulo Freire and the possible dream*. Rotterdam, Sense Publishers.

Bowell, T. 2018. Changing the world one premise at a time: Argument, imagination and post-truth. *In*: Peters, M. A., Rider, S., Hyvönen, M. & Besley, T. (eds.) *Post-truth, fake news: Viral modernity & higher education*. Singapore, Springer.

Cabral, A. 1971. *Our people are our mountains: Amilcar cabral on the Guinean revolution*, Angola and Guiné, London, Committee for Freedom in Mozambique.

Cameron, A. 1989. First mother and the rainbow children. *Healing the wounds*. See Plant 1989.

Campano, G., Honeyford, M. A., Sánchez, L. & Vander Zanden, S. 2010. Ends in themselves: Theorizing the practice of university-school partnering through horizontalidad. *Language Arts*, 87, 277–86.

Carmona, L. G., Whiting, K., Carrasco, A., Sousa, T. & Domingos, T. 2017. Material services with both eyes wide open. *Sustainability*, 1508.

Carnoy, M. 1997. Foreword. *In*: Freire, P. & Freire, A. M. A. (eds.) *Pedagogy of the heart*. New York, Continuum.

Chen, K.-H. 2010. *Asia as method: Toward deimperialization*, Durham, Duke University Press.

Clegg, S. 2011. Cultural capital and agency: Connecting critique and curriculum in higher education. *British Journal of Sociology of Education*, 32, 93–108.

Cobern, W. W. 1994. Comments and criticism. Point: Belief, understanding, and the teaching of evolution. *Journal of Research in Science Teaching*, 31, 583–90.

Code, L. 1984. Toward a "responsibilist" epistemology. *Philosophy and Phenomenological Research*, 45, 29–50.

Connell, R. 2007. *Southern theory: The global dynamics of knowledge in social science*, Cambridge, Polity.

Connell, R. 2013. Using southern theory: Decolonizing social thought in theory, research and application. *Planning Theory*, 13, 210–23.

Connell, R. 2014. Epistemologies of the South: Justice against epistemicide by Boaventura de Sousa Santos. *American Journal of Sociology*, 120, 949–51.

Connell, R., Collyer, F., Maia, J. & Morrell, R. 2017. Toward a global sociology of knowledge: Post-colonial realities and intellectual practices. *International Sociology*, 32, 21–37.

Cook, J., Nuccitelli, D., Green, S. A., Richardson, M., Winkler, B., Painting, R., Way, R., Jacobs, P. & Skuce, A. 2013. Quantifying the consensus on anthropogenic global warming in the scientific literature. *Environmental Research Letters*, 8, 024024.

Cook, J., Oreskes, N., Doran, P. T., Anderegg, W. R. L., Verheggen, B., Maibach, E. W., Carlton, J. S., Lewandowsky, S., Skuce, A. G., Green, S. A., Nuccitelli, D., Jacobs, P., Richardson, M., Winkler, B., Painting, R. & Rice, K. 2016. Consensus on consensus: A synthesis of consensus estimates on human-caused global warming. *Environmental Research Letters*, 11, 048002.

Cudworth, E. 2003. *Environment and society*, New York, Routledge.

D'alisa, G., Demaria, F. & Kallis, G. 2015. *Degrowth: A vocabulary for a new era*, New York; London, Routledge, Taylor & Francis Group.

Dale, R. 2005. Globalisation, knowledge economy and comparative education. *Comparative Education*, 41, 117–49.

Darder, A. (2002). *Reinventing Paulo Feire: A pedagogy of love* (2002 ed.). Boulder, CO: Westview Press.

De Haas, H. 2010. Migration and development: A theoretical perspective1. *International Migration Review (IMR)*, 44, 227–64.

de la Cadena, M. 2015. Uncommoning nature. *eflux journal 56th Venice Biennale*, 22 August 2015.

de Oliveira Andreotti, V. 2011a. The political economy of global citizenship education. *Globalisation, Societies and Education*, 9, 307–10.

de Oliveira Andreotti, V. 2011b. (Towards) decoloniality and diversality in global citizenship education. *Globalisation, Societies and Education*, 9, 381–97.

de Oliveira Andreotti, V., Souza, L. & de Menezes, M. T. 2012. *Postcolonial perspectives on global citizenship education*, New York, Routledge.

de Sousa Santos, B. 2007a. *Another knowledge is possible: Beyond northern epistemologies*, London; New York, Verso.

de Sousa Santos, B. 2007b. Beyond abyssal thinking: From global lines to ecologies of knowledges. *Review (Fernand Braudel Center)*, 30, 45–89.

de Sousa Santos, B. 2008. The World Social Forum and the global left. 36, 2. Oficina de Centro de Estudos Sociais (CES), Coimbra, Portugal, 247–70.

de Sousa Santos, B. 2012a. Public sphere and epistemologies of the South. *Africa Development*, 37, 43–67.

de Sousa Santos, B. 2012b. The university at a crossroads. *Human Architecture: Journal of Sociology of Self-Knowledge*, 10, 7–16.

De Sousa Santos, B. 2014. *Epistemologies of the South: Justice against epistemicide.* Boulder, CO, Paradigm Publishers.

de Sousa Santos, B. 2016. Epistemologies of the South and the future. *From the European South: A Transdisciplinary Journal of Postcolonial Humanities*, Disll, 17–29.

de Sousa Santos, B. 2018. *The end of the cognitive empire: The coming of age of epistemologies of the South*, Durham, Duke University Press.

Derni, A. 2008. The ecolinguistic paradigm: An integrationist trend in language study. *The International Journal of Language Society and Culture*, no. 24, 21–30.

Dewey, J. 1963. *Democracy and education: An introduction to the philosophy of education*, New York, Macmillan.

Doran, P. T. & Zimmerman, M. K. 2009. Examining the scientific consensus on climate change. *Eos, Transactions American Geophysical Union*, 90, 22–3.

Durkheim, E. & Lukes, S. 1982. *The rules of sociological method*, New York, Free Press.

Dussel, E. D. 2012. Transmodernity and interculturality: An interpretation from the perspective of philosophy of liberation. *Transmodernity*, 1, 28–59.

Fanon, F. 1963. *The wretched of the earth*, New York, Grove Press.

Fanon, F. 1967. *Black skin, white masks*, New York, Grove Press.

Figueroa, R. & Harding, S. G. 2003. *Science and other cultures: Issues in philosophies of science and technology*, New York, NY, Routledge.

Fill, A. 2001. Language and ecology: Ecolinguistic perspectives for 2000 and beyond. *In*: Graddol, D. (ed.) *Applied linguistics for the 21st century*. Open University, School of Education, AILA Review.

Findlay, A. M. 2011. Migrant destinations in an era of environmental change. *Global Environmental Change*, 21, S50–8.

Ford, L. 2015. Sustainable development goals: All you need to know. *theguardian.com* [Online]. Available: http://www.theguardian.com/global-development/2015/jan/19/sustainable-development-goals-united-nations (Accessed January 19).

Frankfurt, H. G. 2005. *On bullshit*, Princeton, NJ, Princeton University Press.

Freire, P. 1970. The adult literacy process as cultural action for freedom. *Harvard Educational Review*, 40, 205–25.

Freire, P. 1971. Education as cultural action. *In*: Colonnesse, L. A. (ed.) *Division for Latin America United States Catholic Conference*. Washington, DC.

Freire, P. 1978. *Pedagogy in process: The letters to Guinea-Bissau*, New York, Seabury Press.

Freire, P. 1985. *The politics of education: Culture, power, and liberation*, South Hadley, MA, Bergin & Garvey.

Freire, P. 1992. *Pedagogy of hope*, New York, NY, Continuum.

Freire, P. 1993. *Pedagogy of the city*, New York, Continuum.

Freire, P. 1997. *Pedagogy of the heart*, New York, Continuum.

Freire, P. 1998a. Cultural action and conscientization. *Harvard Educational Review*, 68(4 winter).

Freire, P. 1998b. *Pedagogy of freedom: Ethics, democracy, and civic courage*, Lanham, MD, Rowman & Littlefield.

Freire, P. 1998c. *Politics and education*, Los Angeles, UCLA Latin American Center Publications.

Freire, P. 1998d. *Teachers as cultural workers: Letters to those who dare teach*, Boulder, CO, Westview Press.

Freire, P. 2000. *Pedagogy of the oppressed*, New York, NY, Continuum.

Freire, P. 2004. *Pedagogy of indignation*, Boulder, CO, Paradigm Publishers.

Freire, P. & Macedo, D. P. 1996. *Letters to Cristina: Reflections on my life and work*, New York, Routledge.

Freitas, A. L. S. D. 2012. Conscientization [Conscientização]. *In*: Streck, D. R., Redin, E. & Zitkoski, J. J. (eds.) *Paulo Freire encyclopedia*. Lanham, Rowman & Littlefield Publishers.

Gadotti, M. 1996. *Pedagogy of praxis: A dialectical philosophy of education*, Albany, SUNY Press.

Gadotti, M. 2008a. *Education for sustainability: A critical contribution to the Decade of Education for Sustainable Development*. São Paulo, Brazil: University of São Paulo, Paulo Freire Institute, São Paulo.

Gadotti, M. 2008b. *Education for sustainable development: What we need to learn to save the planet*, São Paulo, Instituto Paulo Freire.

Gadotti, M. 2008c. What we need to learn to save the planet. *Journal of Education for Sustainable Development*, 2, 21–30.

Gadotti, M. & Torres, C. A. 2009. Paulo Freire: Education for development. *Development and Change*, 40, 1255–67.

Giddens, A. 1990. *The consequences of modernity*, Palo Alto, Stanford University Press.

Giddens, A. 1999. *Runaway world: How globalisation is reshaping our lives*, London, Profile.

Giroux, H. A. 2001. *Theory and resistance in education: Towards a pedagogy for the opposition*, Westport, CT, Bergin & Garvey.

Goodman, R. T. & Saltman, K. J. 2002. *Strange love: Or how we learn to stop worrying and love the market*, Lanham, MD, Rowman & Littlefield.

Gore, A. (Dir.) 2006. *An inconvenient truth: A global warning*, Hollywood, CA, Paramount.

Gramsci, A., Hoare, Q. & Nowell-Smith, G. 1972. *Selections from the prison notebooks of Antonio Gramsci*, New York, International Publishers.

Grande, S. 2004. *Red Pedagogy: Native American social and political thought*, Lanham, MD, Rowman & Littlefield Publishers.

Grande, S. 2009. Red pedagogy: Indigenous theories of redistribution (a.k.a sovereignty). *In*: Apple, M. W., Au, W. & Gandin, L. S. A. (eds.) *The Routledge international handbook of critical education*. New York, NY, Routledge.

Grigorov, S. & Fleuri, R. 2012. Ecopedagogy: Educating for a new eco-social intercultural perspective. *Visão Global, Joaçaba*, 15, 433–54.

Grosfoguel, R. 2008. Decolonizing political-economy and postcolonial studies: Transmodernity, border thinking, and global coloniality. *Revista Crítica de Ciências Sociais*, 2008, 115–47.

Gutiérrez, F. & Prado, C. 1989. *Ecopedagogia e cidadania planetária. (Ecopedagogy and planetarian citizenship)*, São Paulo, Cortez.

Gutiérrez, F. & Prado, C. 1999. *Ecopedagogia e cidadania planetaria*, São Paulo, Brazil: Cortez.

Hale, S. 2018. Something resembling "truth": Reflections on critical pedagogy in the new "post-truth" landscape. *Gender and Education Association Annual Conference*. Newcastle, Australia.

Halevi, D. Y. E., Misiaszek, G. W., Kelly, H., Shah, S., Mugarura, C. & Walsh, L. 2018. Building eco-social resilience in the face of climate change and drought: How permaculture pedagogy and praxis can benefit rural communities and their environment. *In*: Leal Filho, W. (ed.) *Climate change management*. London: Springer.

Haluza-Delay, R. 2003. When the topic is racism: Research and advocacy with a community coalition. *Social Justice*, 30(4) 77–90.

Hannigan, J. A. 2006. *Environmental sociology*, London, Routledge.

Harding, S. G. 1991. *Whose science? Whose knowledge?: Thinking from women's lives*, Ithaca, Cornell University Press.

Harding, S. G. 1998. *Is science multicultural?: Postcolonialisms, feminisms, and epistemologies*, Bloomington, IN, Indiana University Press.

Harding, S. G. 2004. *The feminist standpoint theory reader: Intellectual and political controversies*, New York, Routledge.

Harding, S. G. 2006. *Science and social inequality: Feminist and postcolonial issues*, Urbana, University of Illinois Press.

Harding, S. G. & Hintikka, M. B. 1983. *Discovering reality: Feminist perspectives on epistemology, metaphysics, methodology, and philosophy of science*, Dordrecht, Holland, D. Reidel.

Harris, I. M. & Morrison, M. L. 2003. *Peace education*, Jefferson, NC, McFarland.

Hartsock, N. C. M. 1998. *The feminist standpoint revisited and other essays*, Boulder, CO, Westview Press.

Held, D. 1980. Introduction to critical theory: Horkheimer to Habermas. *In*: Held, D. (ed.) *Introduction to critical theory*. Berkeley, CA: University of Southern California Press.

Horton, M., Freire, P., Bell, B., Gaventa, J. & Peters, J. M. 1990. *We make the road by walking: Conversations on education and social change*, Philadelphia, Temple University Press.

Horton, M., Kohl, J. & Kohl, H. R. 1998. *The long haul: An autobiography*, New York, Teachers College Press.

Illich, I. 1983. *Deschooling society*, New York, Harper Colophon.

International Organization for Migration (IOM) 2014. *IOM outlook on migration, environment and climate change*. Geneva: International Organization for Migration.

Jasanoff, S. & Simmet, H. R. 2017. No funeral bells: Public reason in a "post-truth" age. *Social Studies of Science*, 47, 751–70.

Jickling, B. & Wals, A. E. J. 2008. Globalization and environmental education: Looking beyond sustainable development. *Journal of Curriculum Studies*, 40, 1–21.

Kahn, R. 2010. *Critical pedagogy, ecoliteracy, and planetary crisis: The ecopedagogy movement*, New York, Peter Lang.

Kahn, R. & Kellner, D. M. 2006. Oppositional politics and the internet: A critical/reconstructive approach. *In*: Durham, M. G. & Kellner, D. (eds.) *Media and cultural studies: Keyworks*. Rev. ed. Malden, MA: Blackwell.

Kallis, G. & March, H. 2015. Imaginaries of hope: The utopianism of degrowth. *Annals of the Association of American Geographers*, 105, 360–8.

Katz, M. B. 2008. *The price of citizenship: Redefining the American welfare state*, Philadelphia, University of Pennsylvania Press.

Kellner, D. 2002. Theorizing globalization. *Sociological Theory*, 20, 285–305.

King, J. E. 2017. 2015 AERA Presidential Address: Morally engaged teach/ers dismantling epistemological nihilation in the age of impunity. *Educational Researcher*, 46, 211–22.

Kinsman, G. 2006. Mapping social relations of struggle: Activism, ethnography, social organization. *In*: Frampton, C. (ed.) *Sociology for changing the world: Social movements/social research*. Black Point, NS, Fernwood.

Klees, S. J. 2008a. Neoliberalism and education revisited. *Globalisation, Societies and Education*, 6, 409–14.

Klees, S. J. 2008b. A quarter century of neoliberal thinking in education: Misleading analyses and failed policies. *Globalisation, Societies and Education*, 6, 311–48.

Kolinjivadi, V. 2019. Avoiding dualisms in ecological economics: Towards a dialectically-informed understanding of co-produced socionatures. *Ecological Economics*, 163, 32–41.

Kuhn, T. S. 1970. *The structure of scientific revolutions*, Chicago, University of Chicago Press.

Ladson-Billings, G. 2009. Race still matters: Critical race theory in education. *In*: Apple, M. W., Au, W. & Gandin, L. S. A. (eds.) *The Routledge international handbook of critical education*. New York, NY: Routledge.

Lahar, S. 1991. Ecofeminist theory and grassroots politics. *Hypatia*, 6, 28–45.

Lather, P. 1986. Issues of validity in openly ideological research: Between a rock and a soft place. *Interchange*, 17, 63–84.

Latouche, S. 2004. Degrowth economics. *Le Monde Diplomatique*, 11, 2004.

Lema, E., Mbilinyi, M. & Rajani, R. 2004. *Nyerere on Education (Nyerere kuhusu Elimu): Selected essays and speeches (1954 -1998)*, Dar es Salaam, Tanzania, HakiElimu.

Leopold, A. & Brooks, A. 1933. *Game management*, New York, London, C. Scribner's Sons.

Lima, L. C. 2007. Schooling for critical education: The reinvention of schools as democratic organizations. *In*: Torres, C. A. & Teodoro, A. (eds.) *Critique and utopia: New developments in the sociology of education in the twenty-first century*. Lanham, MD: Rowman & Littlefield.

Lister, M. 2014. Climate change refugees. *Critical Review of International Social and Political Philosophy*, 17, 618–34.

Marquand, D. 1991. *The progressive dilemma*, London, Heinemann.

Marshall, H. 2011. Instrumentalism, ideals and imaginaries: Theorising the contested space of global citizenship education in schools. *Globalisation, Societies and Education*, 9, 411–26.

Martínez-Alier, J. 2012. Environmental justice and economic degrowth: An alliance between two movements. *Capitalism Nature Socialism*, 23, 51–73.

Marx, K. 1977. *Capital: Volume one*, New York, Vintage Books.

Max-Neef, M. A. 2005. Foundations of transdisciplinarity. *Ecological Economics*, 53, 5–16.

Mcgregor, J. 1993. Refugees and the environment. *In*: Black, R. & Robinson, V. (eds.) *Geography and refugees: Patterns and processes of change*. New York, Belhaven Press.

Mcintyre, L. C. 2018. *Post-truth*, Cambridge, MA, MIT Press.

Mclaren, P. 2007. Conservation, class struggle, or both: A response to C.A. Bowers. *Capitalism, Nature, and Socialism*, 18, 99–108.

Mcnamara, K. E. 2007. Conceptualizing discourses on environmental refugees at the United Nations. *Population and Environment*, 29, 12–24.

Melosi, M. V. 1995. Equity, eco-racism and environmental history. *Environmental History Review*, 19, 1–16.

Memmi, A. 1967. *The colonizer and the colonized*, Boston, Beacon Press.

Memmi, A. 1991a. Chapter 2: Portrait of the colonized: situations of the colonized. *In*: Memmi, A. (ed.) *The colonizer and the colonized*. Expanded ed. Boston: Beacon Press.

Memmi, A. 1991b. *The colonizer and the colonized*, Boston, Beacon Press.

Merchant, C. 2003. Shades of darkness: Race and environmental history. *Environmental History*, 8, 380–94.

Meredith, M. 2007. *Diamonds, gold, and war: The British, the Boers, and the making of South Africa*, New York, PublicAffairs.

Misiaszek, G. W. 2011. *Ecopedagogy in the age of globalization: Educators' perspectives of environmental education programs in the Americas which incorporate social justice models*. Ph.D., University of California, Los Angeles.

Misiaszek, G. W. 2012. Transformative environmental education within social justice models: Lessons from comparing adult ecopedagogy within North and South America. *In*: Aspin, D. N., Chapman, J., Evans, K. & Bagnall, R. (eds.) *Second international handbook of lifelong learning*. London, Springer.

Misiaszek, G. W. 2014a. Ecopedagogy and comparative education. *Newsletter of the World Council of Comparative Education Societies (WCCES)*, 1, 11–14.

Misiaszek, G. W. 2014b. Environmental education through global and local lenses: Ecopedagogy and globalizations in Appalachia, Argentina, and Brazil. *In*: Turner, D. A. & Yolcu, H. (eds.) *Neoliberal educational reforms: A critical analysis*. New York: Taylor & Francis/Routledge.

Misiaszek, G. W. 2014c. Learning from Southern environmental education models: Borrowing ecopedagogy through processes of globalization from below. *Éducation Comparée*, 209–41.

Misiaszek, G. W. 2015. Ecopedagogy and citizenship in the age of globalisation: Connections between environmental and global citizenship education to save the planet. *European Journal of Education*, 50, 280–92.

Misiaszek, G. W. 2016a. Ecopedagogy as an element of citizenship education: The dialectic of global/local spheres of citizenship and critical environmental pedagogies. *International Review of Education*, 62, 587–607.

Misiaszek, G. W. 2016b. Ecopedagogy as an element of citizenship education: The dialectic of global/local spheres of citizenship and critical environmental pedagogies in the Americas. *International Review of Education - Journal of Lifelong Learning (UNESCO)*, 62 (5).

Misiaszek, G. W. 2018a. A critical scholar's journey in China: A brief Freirean analysis of insider–outsider tensions. *Educational Philosophy and Theory*, 50 (12), 1133–43.

Misiaszek, G. W. 2018b. *Educating the global environmental citizen: Understanding ecopedagogy in local and global contexts*, New York, Routledge.

Misiaszek, G. W. 2019. The end of the cognitive empire: The coming of age of epistemologies of the south by Boaventura de Sousa Santos. *Comparative Education Review*, 63, 452–4.

Misiaszek, G. W. 2020. Ecopedagogy: Teaching critical literacies of "development", "sustainability", and "sustainable development". *Teaching in Higher Education*, 25(5), 615–32. doi:10.1080/13562517.2019.1586668

Misiaszek, G. W. 2020a. Countering post-truths through ecopedagogical literacies: Teaching to critically read "development" and "sustainable development". *Educational Philosophy & Theory* 52 (7).

Misiaszek, G. W. 2020b. Locating and diversifying modernity: Deconstructing knowledges to counter development for a few. In: Peters, M. A., Besley, T., Jandrić, P. & Zhu, X. (eds.) *Knowledge socialism: The rise of peer production: Collegiality, collaboration, and collective intelligence*. Singapore, Springer.

Misiaszek, G. W. & Misiaszek, L. I. 2016. Global citizenship education and ecopedagogy at the intersections: Asian Perspectives in comparison. *Asian Journal of Education*, 17, 11–37.

Misiaszek, G. W. & Torres, C. A. 2019. Ecopedagogy: The missing chapter of pedagogy of the oppressed. *In*: Torres, C. A. (ed.) *Wiley handbook of Paulo Freire*. Hoboken, NJ: Wiley-Blackwell.

Morrow, R. A. 2006. Forward - Critical theory, globalization, and higher education: Political economy and the cul-de-sac of the postmodernist cultural turn. *In*: Rhoads, R. A. & Torres, C. A. (eds.) *The university, state, and market: The political economy of globalization in the Americas*. Stanford: Stanford University Press.

Morrow, R. A. 2019. Paulo Freire and the "logic of reinvention": Power, the State, and education in the global age. *In*: Torres, C., A. (ed.) *Wiley handbook of Paulo Freire*. Hoboken, NJ: Wiley-Blackwell.

Naess, A. 1990. Deep ecology. *In*: Dobson, A. (ed.) *The green reader*. London, André Deutsch.

O'cadiz, M. P., Wong, P. & Torres, C. A. 1998. *Education and democracy: Paulo Freire, social movements, and educational reform in São Paulo*, Boulder, Westview Press.

Olmos, L. & Torres, C. A. 2009. Theories of the state, educational expansion, development, and globalizations: Marxian and critical approaches. *In*: Cowen, R. & Kazamias, A. M. (eds.) *International handbook of comparative education*. Dordrecht: Springer.

Oreskes, N. 2004. The scientific consensus on climate change. *Science*, 306, 1686–6.

Orr, D. W. 2009. *Down to the wire: Confronting climate collapse*, Oxford; New York, Oxford University Press.

Park, J. 2017. Knowledge production with Asia-Centric research methodology. *Comparative Education Review*, 61, 760–79.

Peters, M. A. 2017. Education in a post-truth world. *Educational Philosophy and Theory*, 49, 563–6.

Peters, M. A., Rider, S., Hyvönen, M. & Besley, T. 2018. *Post-truth, fake news: Viral modernity & higher education*, Singapore, Springer.

Piguet, E. 2013. From “Primitive Migration” to “Climate Refugees”: The Curious Fate of the Natural Environment in Migration Studies. *Annals of the Association of American Geographers*, 103, 148–62.

Pomeroy, D. 1994. Science education and cultural diversity: Mapping the field. *Studies in Science Education*, 24, 49–73.

Pongratz, L. 2005. Critical theory and pedagogy: Theodor W. Adorno and Max Horkheimer’s contemporary significance for a critical pedagogy. *In*: Fischman, G., Mclaren, P., Sunker, H. & Lankshear, C. (eds.) *Critical theories, radical pedagogies, and global conflicts*. Lanham, MD: Rowman & Littlefield Publishers.

Post, D. & Meng, Y. 2018. Does schooling foster environmental values and action? A cross-national study of priorities and behaviors. *International Journal of Educational Development*, 60, 10–18.

Postma, D. W. 2006. *Why care for nature?: In search of an ethical framework for environmental responsibility and education*, New York, Springer.

Raworth, K. 2017. *Doughnut economics: Seven ways to think like a 21st century economist*, White River Junction, Vermont, Chelsea Green Publishing.

Rhoads, R. A. & Torres, C. A. 2006. *The university, state, and market the political economy of globalization in the Americas*, Stanford, CA, Stanford University Press.

Rice, A. 2016. This is New York in the not-so-distant future. *New York Magazine* [Online]. Available: http://nymag.com/intelligencer/2016/09/new-york-future-flooding-climate-change.html (Accessed December 1, 2019).

Richardson, G. 2008. Caught between imaginaries: Global citizenship and the persistence of the nation. *In*: Abdi, A. A. & Shultz, L. (eds.) *Educating for human rights and global citizenship*. Albany: State University of New York Press.

Rizvi, F. 2006. Imagination and the globalisation of educational policy research. *Globalisation, Societies and Education*, 4, 193–205.

Said, E. W. 1979. *Orientalism*, New York, Vintage Books.

Sarukkai, S. 2012. Experience and theory: From Habermas to Gopal Guru. *In*: Guru, G. (ed.) *The Cracked Mirror: An Indian Debate on Experience and Theory*. Oxford, UK, Oxford University Press.

Saul, J. R. 1995. *The unconscious civilization*, Concord, ON, House of Anansi Press.

Savelyeva, T. 2017. Vernadsky meets Yulgok: A non-Western dialog on sustainability. *Educational Philosophy and Theory*, 49, 501–20.

Scherr, A. 2005. Social subjectivity and mutual recognition as basic terms of a critical theory of education. *In*: Fischman, G., Mclaren, P., Sunker, H. & Lankshear, C. (eds.) *Critical theories, radical pedagogies, and global conflicts*. Lanham, MD: Rowman & Littlefield Publishers.

Schneider, F., Kallis, G. & Martinez-Alier, J. 2010. Crisis or opportunity? Economic degrowth for social equity and ecological sustainability. Introduction to this special issue. *Journal of Cleaner Production*, 18, 511–18.

Shiva, V. 1992. Recovering the real meaning of sustainability. *In*: Cooper, D. E. & Palmer, J. (eds.) *The Environment in question: Ethics and global issues*. London: Routledge.

Shultz, L. 2007. Educating for global citizenship: Conflicting agendas and understandings. *The Alberta Journal of Educational Research*, 53, 248–58.

Shultz, L. & Jorgenson, S. 2009. *Global citizenship education in post-secondary institutions: A review of the literature*. Edmonton: University of Alberta, 10, 2014.

Singh, N. M. 2019. Environmental justice, degrowth and post-capitalist futures. *Ecological Economics*, 163, 138–42.

Smith, G. H. 2000. Protecting and respecting indigenous knowledge. *In*: Battiste, M. (ed.) *Reclaiming indigenous voice and vision*. Vancouver: University of British Columbia Press.

Smith, G. H. 2015. Transforming research: The Indigenous struggle for social, cultural, and economic justice within and through education. *American Educational Research Association Annual Meeting*, April 9, Chicago, Illinois.

Smith, L. T. 1999. *Decolonizing methodologies: Research and indigenous peoples*, London, Zed Books.

Smith, L. T. 2005. On tricky ground: Researching the native in the age of uncertainty. *In*: Denzin, N. K. & Lincoln, Y. S. (eds.) *The SAGE handbook of qualitative research*. 3rd ed. Thousand Oaks, CA: Sage Publications.

Snively, G. & Corsiglia, J. 2001. Discovering indigenous science: Implications for science education. *Science Education*, 85, 6–34.

Sontag, S. 1977. *On photography*, New York, Dell Publishing Co.

Spash, C. L. 2013. The shallow or the deep ecological economics movement? *Ecological Economics*, 93, 351–62.

Spivak, G. C. 1988. *Can the subalterns speak?* Basingstoke, Macmillan.

Spivak, G. C. & Harasym, S. 1990. *The post-colonial critic: Interviews, strategies, dialogues*, New York; London, Routledge.

Stapp, W., Benett, D., Bryan, W., Fulton, J., MacGregor, J., Nowak, P., . . . Avlick, S. (1969). The concept of environmental education. *Journal of Environmental Education*, 1(1), 30–1.

Stein, S. 2017. The persistent challenges of addressing epistemic dominance in higher education: Considering the case of curriculum internationalization. *Comparative Education Review*, 61, S25–50.

Stibbe, A. 2012. Ecolinguistics and globalization. *In: The handbook of language and globalization*. West Sussex, UK, Wiley-Blackwell.

Stibbe, A. 2014. An ecolinguistic approach to critical discourse studies. *Critical Discourse Studies*, 11, 117–28.

Stornaiuolo, A., Smith, A. & Phillips, N. C. 2017. Developing a transliteracies framework for a connected world. *Journal of Literacy Research*, 49, 68–91.

Stromquist, N. P. 2002a. Chapter 1: Theorizing globalization. *In*: Stromquist, N. P. (ed.) *Education in a globalized world: The connectivity of economic power, technology, and knowledge*. Lanham, MD: Rowman & Littlefield Publishers.

Stromquist, N. P. 2002b. *Education in a globalized world: The connectivity of economic power, technology, and knowledge*, Lanham, MD, Rowman & Littlefield Publishers.

Suarez, D. F. & Ramirez, F. O. 2007. Human rights and citizenship: The emergence of human rights education. *In*: Torres, C. A. & Teodoro, A. (eds.) *Critique and utopia: New developments in the sociology of education in the twenty-first century*. Lanham, MD: Rowman & Littlefield.

Swyngedouw, E. 1996. The city as a hybrid: On nature, society and cyborg urbanization. *Capitalism Nature Socialism*, 7, 65–80.

Syed, N. I. & Misiaszek, G. W. 2020. Unmentionable GCE? Teaching global citizenship as dis/empowering environmentalism and sustainable "development" in India. *In*: Misiaszek, L. I. (ed.) *Exploring the complexities in global citizenship education: Hard spaces, methodologies, and ethics*. New York: Routledge.

Takayama, K. 2009. Progressive education and critical education scholarship in Japan: Toward the democratization of critical educational studies. *In*: Apple, M. W., Au, W. & Gandin, L. S. A. (eds.) *The Routledge international handbook of critical education*. New York, NY: Routledge.

Takayama, K., Sriprakash, A. & Connell, R. 2017. Toward a postcolonial comparative and international education. *Comparative Education Review*, 61, S1–S24.

Tarozzi, M. & Torres, C. A. 2016. *Global citizenship education and the crises of multiculturalism: Comparative perspectives*, London, Bloomsbury Academic.

Taylor, A. 2014. Bhopal: The world's worst industrial disaster, 30 years later. *The Atlantic* [Online]. Available: https://www.theatlantic.com/photo/2014/12/bhopal-the-worlds-worst-industrial-disaster-30-years-later/100864/ (Accessed December 1, 2019).

Teodoro, A. & Torres, C. A. 2007. Introduction: Critique and utopia in the sociology of education. *In*: Torres, C. A. & Teodoro, A. (eds.) *Critique and utopia: New developments in the sociology of education in the twenty-first century*. Lanham, MD: Rowman & Littlefield Publications.

Tierney, R. J. 2017a. Global compliance and the challenges of internationalizing educational research in the age of the Western empire building. *Sixth Worldwide Forum for Comparative Education*. Beijing Normal University, Beijing, China.

Tierney, R. J. 2017b. Multitopia: Global citizenship across diverse spaces. *Global Commons Review*, 1, 75–81.

Tierney, R. J. 2017c. Populism and literacy consciousness in uncertain global times. *Journal of Literacy Research*, 305–8.

Tierney, R. J. 2018a. Global educational research in Western times: The rise and plight of Chinese educational research. *Frontiers of Education in China*, 13, 163–92.

Tierney, R. J. 2018b. Toward a model of global meaning making. *Journal of Literacy Research*, 50 (4), 397–422.

Torres, C. A. 1998a. Democracy, education, and multiculturalism: Dilemmas of citizenship in a global world. *Comparative Education Review*, 42, 421–47.

Torres, C. A. 1998b. *Democracy, education, and multiculturalism: Dilemmas of citizenship in a global world*, Lanham, MD, Rowman & Littlefield Publishers.

Torres, C. A. 2002. Globalization, education, and citizenship: Solidarity versus markets? *American Educational Research Journal*, 39, 363–78.

Torres, C. A. 2003. Globalizations and education. *Meeting of the Fondazione Liberal.* Milan, Italy.

Torres, C. A. 2009. *Globalizations and education: Collected essays on class, race, gender, and the state*, New York, Teachers College Press.

Torres, C. A. 2011. Public universities and the neoliberal common sense: Seven iconoclastic theses. *International Studies in Sociology of Education*, 21, 177–97.

Torres, C. A. 2013a. Afterword: Another globalization is possible. *In*: Arnove, R. F., Torres, C. A. & Franz, S. (eds.) *Comparative education: The dialectic of the global and the local.* 4th ed. Lanham, MD: Rowman & Littlefield.

Torres, C. A. 2013b. Emerging global citizenship education agenda. *Newsletter of the World Council of Comparative Education Societies (WCCES)*, 2, 11–3.

Torres, C. A. 2014. *First Freire: Early writings in social justice education*, New York, Teachers College Press.

Torres, C. A. 2016. Education for global citizenship. *Oxford Research Encyclopedia of Education.* Oxford, UK: Oxford University Press.

Torres, C. A. 2017. *Theoretical and empirical foundations of critical global citizenship education*, New York, Routledge.

Torres, C. A. (ed.) 2019. *The Wiley handbook of Paulo Freire*, Hoboken, NJ, Wiley-Blackwell.

Torres, C. A. & Rhoads, R. A. 2006. Introduction: Globalization and higher education in the Americas. *In*: Rhoads, R. A. & Torres, C. A. (eds.) *The university, state, and market: The political economy of globalization in the Americas.* Stanford: Stanford University Press.

Twine, R. T. 2001. Ecofeminisms in process. *Ecofeminism e-journal* [Online]. Available: https://www.scribd.com/document/59622755/Ecofeminisms-in-Process (Accessed August 8, 2017).

UNESCO 2017. *Education for sustainable development goals: Learning objectives*, Paris, United Nations Educational, Scientific and Cultural.

Vargas, V. 2003. Feminism, globalization and the global justice and solidarity movement. *Cultural Studies*, 17, 905–20.

Warren, K. J. Environmental stewardship: An ecofeminist philosophical perspective. Environmental stewardship: History, theory and practice workshop proceedings (March 11–12, 1994), 1994. 42.

Warren, K. J. 2000. *Ecofeminist philosophy: A Western perspective on what it is and why it matters*, Lanham, MD, Rowman & Littlefield Publications.

Warren, K. J. & Cheney, J. 1991. Ecological feminism and ecosystem ecology. *Hypatia*, 6, 179–97.

White, H. V. 1973. *Metahistory: The historical imagination in nineteenth-century Europe*, Baltimore, Johns Hopkins University Press.

White, H. V. 1987. *The content of the form: Narrative discourse and historical representation*, Baltimore, John Hopkins University Press.

Whiting, K., Konstantakos, L., Misiaszek, G. W., Simpson, E. & Carmona, L. 2018. Education for the sustainable global citizen: What can we learn from stoic philosophy and Freirean environmental pedagogies? *Education Sciences*, 8 (4), 204.

Zhao, J., Beckett, G. H. & Wang, L. L. 2017. Evaluating the research quality of education journals in China: Implications for increasing global impact in peripheral countries. *Review of Educational Research*, 87, 583–618.

Index

Anthropo-Bio-Centric Gaps, *see* world-Earth *under* distancing
anthropocentric, *see also* world-Earth
  vs. planetary 6–7
anti-coloniality, *see* coloniality
anti-globalization, *see* globalizations
authentic dialogue, *see under* Freire, Paulo

Beck, Ulrich 147
Bill and Melinda Gates Foundation 60–1, *see also* NGOs
bullshit 55–6, 229–42

citizenships
  coloniality 212–13 (*see also entry*)
  contested terrain 204–6
  development (*see under entry*)
  economics 207–11
  ecopedagogical literacies 225, 201–25 (*see also entry*)
  Freire, Paulo 222
  Global Citizenship Education (GCE) (*see entry*)
  globalizations 213–20
  livelihood 221–5 (*see also under* development)
  othering 76–7, 150, 209–11
  Planetary Citizenship Education (PCE) (*see entry*)
  pluralization of citizenship spheres 24, 76–7, 204–11
  solidarity 205–7, 209–25
  with ecopedagogy (introduction) 4–5
  world-Earth (*see entry*)
climate change, *see* global warming
coloniality 51, 165–9, 182
  All Rhodes Must Fall (*see* South Africa *under* environmental problems)
  citizenships (*see under entry*)
  development (*see under entry*)
  Franz Fanon (*see entry*)
  Memmi, Albert (*see entry*)
  myth (countering) 165–7
  neocoloniality (*see entry*)
  Said, Edward (*see entry*)
common sense 154, 185–6, 193–4, 225
comparative and international education (CIE) 10–11, 38, 64, 145, 201–2
conflict, *see under* peace education
consciousization (i.e., *conscientização*), *see under* Freire, Paulo
consumption, *see under* environmental problems
continuum of environmental violence, *see under* development
corporations, *see* transnational corporations (TNCs)
critical race theories, *see under* theorizing
critical theories, *see* theorizing
curricula
  absences/emergences of disciplines (*see under* ecopedagogy)
  hidden curricula (*see entry*)
  problematizing 116–17
  transdisciplinary (*see under entry*)

Dante (Durante degli Alighieri) 143
de-coloniality, *see* coloniality
de-development, *see* de-growth *under* development
deep ocean oil drilling, *see under* environmental problems
de-growth, *see under* development; Singh, Neera M.
democratic
  development (*see under entry*)
  education (*see* ecopedagogy; Freire, Paulo)
  thickness and thinness 203–4
de Sousa Santos, Boaventura 27, 41, 50–5, 70–1, 81–2, 163–4, 167–8, 177–8, 181–8
  absences/emergences of disciplines (*see under* ecopedagogy)
  absences/emergences of sociologies 74–5, 189–90

ecologies of knowledges (*see under* epistemologies)
epistemicide 108–9, 179–80
epistemologies of the South/North (*see under* epistemologies)
globalizations 57–8, 108–9
histories of disciplines 159–60
postabyssal pedagogies 163–4
self-reflectivity limitations 133, 147–8, 164, 186–8, 207 (*see also* limitations of reflectivity *under* ecopedagogy)
solidarity 205–7
thicker experiences 195
development
authentic development 82–3, 168–9
citizenships 22–3, 75–90, 202–4, 211–21
coloniality 83, 168–9, 218
consumption (*see under* environmental problems)
continuum of environmental violence 19, 75–6
de-development (*see* de-growth *under this entry*)
de-growth 77–9
democratic 83, 203–4
epistemologies 180–1, 184, 191–3
evolving (*vs.* development/transformation) 98–101
Freire, Paulo (*see under entry*)
futures (*see under entry*)
globalizations (*see under entry*)
Hegel's Master-Slave dialectic (*see entry*)
hope *vs.* fatalism 102–3 (*see also* hope *and* utopia and education *under* Freire, Paulo)
livelihood 69, 75–6, 88–90, 126–7, 149–50, 162, 202–3, 209–10, 211–12, 221–5, 231–2, 240 (*see also under* citizenships)
neoliberalism (*see under entry*)
post-truthism (*see under entry*)
reading "development" (*see* development (introduction) *under* ecopedagogical literacies)
reading "sustainability" (*see* sustainability (introduction) *under* ecopedagogical literacies)
Sustainable Development Goals (SDGs) (*see entry*)
technologies (*see* sciences and technologies)
utopia and unfinishness (*see* Freire, Paulo)
zero-sum goals 88, 209, 233
Dewey, John 4, 83–4, 124–5, 204
dialogue/dialectics, *see* dialogue *under* Freire, Paulo
disciplines, *see* transdisciplinary
distancing
economics 71–2
globalizations 63–5
Organic Ecopedagogical Modeling (OEM) (*see entry*)
world-Earth 6–7, 19–21, 32–6, 144–5, 157, 205–6
Durkheim, Émile 38

Earth-world, *see* world-Earth
ecofeminism, *see under* feminism
ecolinguistics 18–19
ecologies of knowledges, *see under* epistemologies
economics 1
citizenships (*see entry*)
consumption (*see under* environmental problems)
distancing (*see under entry*)
globalizations 69–70
justice (economic justice) 69–70
labor (*see* entry)
neoliberalism (*see under entry*)
ecopedagogical literacies 1–2, 18, 44–89
deepening and widening 25–9, 38–41, 48–9, 58–9, 145–6, 163–4, 246
development (introduction) 2–3, 5–6, 11–12
epistemologies (*see* epistemologies)
Freirean literacy (*see under* Freire, Paulo)
global meaning making (*see* Tierney, Robert J.)
media (*see entry*)
Organic Ecopedagogical Modeling (OEM) (*see entry*)
positionality (*see* entry)

reading world-Earth 17–22, 111–13, 178–9 (*see also* world-Earth; world-Earth *under* distancing)
sustainability (introduction) 22–3, 79–80, 84–5
ecopedagogy
absences/emergences of disciplines 74–5, 189–90
curricula (*see entry*)
de-distancing (*see* distancing)
deepening and widening (*see under* ecopedagogical literacies)
Freirean reinvention (*see* Freire, Paulo)
Freire's ecopedagogical work (*see* ecopedagogical work *under* Freire, Paulo)
hope (*see under* Freire, Paulo)
intersectionality with other environmental pedagogies 31–2
key questions 7–8
limitations of reflectivity 37–42, 75–6, 117–18, 145–7, 152, 186–8
literacy (*see* ecopedagogical literacies)
natural disasters 21–2
plurality of ecopedagogies 31–2
praxis (*see entry*)
problem-posing 45–7, 80–1
radical essence 129–31, 134–5, 152–72
research (*see entry*)
research spaces within ecopedagogical spaces (*see under* research)
ecoracism, *see* critical race theories *under* theorizing
education for sustainable development (ESD) 83–90, 191–3
d/*D*evelopment 22–3, 83–8
environmental education, *see* environmental pedagogies
environmental migration, *see under* environmental problems
environmental movements, *see* social-environmental movements
environmental pedagogies
education for sustainable development (ESD) (*see entry*)
intersectionality with ecopedagogies (*see under* ecopedagogy)
pedagogies on the environment 17–18, 31, 240
why fail (shallow models) 9–10, 109–10, 157–9
environmental problems
air pollution 147
climate change (*see* global warming)
consumption 19, 89, 127, 149–53, 211, 222, 231 (*see also* neoliberalism; Westernization)
continuum of environmental violence (*see under* development)
deep ocean oil drilling (*see entry*)
deep oil drilling 19
deforestation 62
environmental migration 89–90, 244–6
ewaste 214
global warming (*see entry*)
hypocrisy 127–8, 149
industrial *F*arming 131–2
mining 166–7, 214–15 (*see also* mountain top removal (MTR) *and* South Africa *under this entry*)
mountain top removal (MTR) 75
natural disasters (*see under* ecopedagogy)
non-democratic 204
nuclear energy 56
nuclear war 143
South Africa 166–7
epistemologies 24–5, 47–56, 177–99
deepening and widening 46–7, 65–6, 138–9, 144–5 (*see also under* ecopedagogical literacies)
de Sousa Santos, Boaventura (*see entry*)
development (*see under entry*)
dominance of Nature (*see under* Nature; world-Earth)
ecologies of knowledges 50–1, 81–2, 117–18, 179–80
epistemologies of the South/North 50–2, 181–8
global meaning making (*see* Tierney, Robert J.)
histories of disciplines 159–60 (*see also* absences/emergences of disciplines *under* ecopedagogy)
knowledge legitimization 52–3, 114–15, 177–9
Organic Ecopedagogical Modeling (OEM) (*see entry*)

positionality (*see entry*)
post-truthism 240–1 (*see also entry*)
Western individualism (*see under* Westernization)
epistemologies of the South/North, *see under* epistemologies
evolving, *see* evolving (*vs.* development/transformation) *under* development

Fanon, Franz 50–1, 106, 148, 167–8, 212–13
feminism 31–2, 145, 182
ecofeminism 31, 169–72, 231
quilting theories 169–72
stand-point theory 46–7
Frankfurt, Harry, *see* bullshit
Freire, Paulo 95–140
anti-environmentalist myths 124–5, 130–1
authentic authority 29–30
authentic dialogue 115–16, 119–21, 128
banking education 29–30, 44–5, 100, 109–14, 197–8
bottom-up teaching and methodologies 131–5
citizenships (*see under entry*)
class suicide 87–8, 148–50, 224–5
consciousization (i.e., conscientização) 47, 124, 134
cultural circles 123
democratic and safe learning spaces 29–30, 131–2
development 217–19
dialectical 28–30, 41–2, 113–23, 138–9, 187–8 (*see also* authentic dialogue *under this entry*)
ecopedagogical work 33–5, 96–8
fatalism (*see* utopia and education *under this entry*)
generative themes 36–7, 42, 145
globalizations 104–9, 129–30, 158
Hegel's Master-Slave dialect (*see entry*)
hope 102–4, 143–4, 143–4 (*see also* utopia and education *under this entry*)
human uniqueness 36, 79–80, 83–4, 95–6, 106, 118, 127, 196–8 (*see also* world-Earth)
limit situations 61, 103–4, 136–40, 162–3, 197–9, 229–46
literacy 18, 44–5 (*see also* ecopedagogical literacies)
love 215–17
neoliberalism (*see under entry*)
oppressor-oppressed positionality (*see under* positionality)
politics 47–56, 121
praxis (*see under entry*)
problem-posing 25–7
reinvention 95–8, 132
research as curiosity (*see under* research)
revolution 129
sciences and technologies (*see under entry*)
"success" problematizing 115, 125–7
thematic universe 27, 36–7, 136–8, 145
transformation (*see entry; evolving* (*vs.* development/transformation) *under* development)
unfinishedness 25, 85–6, 98–102, 106–7 (*see also* futures)
universal ethic 215–17
utopia and education 22, 35–6, 73–4, 105–7, 104–9, 138, 152 (*see also* futures)
world-Earth (*see entry*; futures)
futures
development 103–4, 128–9
Freire, Paulo 100–1, 103–4, 128–9

Gadotti, Moacir 23–5, 82–3, 96–7, 123–4, 134, 201–2, 204
generative themes, *see under* Freire, Paulo
Giddens, Anthony 58–60
Global Citizenship Education (GCE)
contested terrain 204–6
global commons 214–17, 219–20
globalizations 57, 62–3
neoliberal GCE 207–11
global commons, *see under* Global Citizenship Education (GCE)
globalizations
agents 59–60
citizenships (*see under entry and Global Citizenship Education (GCE)*)
contested terrain 49, 56–66, 106
de Sousa Santos, Boaventura (*see under entry*)

development 61–8, 155, 213–20
distancing (*see under entry*)
economics (*see under entry*; neoliberalism)
Freire, Paulo (*see under entry*)
Giddens, Anthony (*see entry*)
Global Citizenship Education (GCE) (*see entry*)
global meaning making (*see* Tierney, Robert J.)
media (*see under entry*)
neocoloniality (*see under entry*)
neoliberalism (*see under entry*)
NGOs (Non-Governmental Organizations) (*see under entry*)
Planetary Citizenship Education (PCE) (*see under entry*)
post-truthism (*see entry*)
Reaganism and Thatcherism (*see entry*)
social media (*see* media)
transnational corporations (TNCs) (*see under entry*)
Westernization (*see entry*)
World Bank (*see entry*)
global meaning making, *see* Tierney, Robert J.
global warming 6–8, 38, 72, 87, 128–30, 233–4, 238
Google, *see under* media
Gramsci, Antonio 191–2

Hale, Sondra 230–1
Hegel's Master-Slave dialectic 131, 133–5, 153, 220
hidden curricula 17–18, 60–1, 70
Highlander Research and Education Center (Highlander Folk School), *see* Horton, Myles
Horton, Myles 10, 81, 121–3

identities, *see* human uniqueness *and* thematic universe *under* Freire, Paulo
Illich, Ivan 32, 75–6, 79–80, 126, 225
interdisciplinary, *see* transdisciplinary
international education, *see* comparative and international education (CIE)

knowledges, *see* epistemologies
Kuhn, Thomas 229

labor 71–2, 124–5, 158, 166, 214, 222
limit situations, *see under* Freire, Paulo
linguistics, *see* ecolinguistics
livelihood, *see under* citizenships; development

Marx, Karl 25, 117–18, 125, 132, 142–3, 152, 229–30, 235
Maslow Hierarchy of Needs 152–3, 221
media 238
critical media culture theories 60–1, 233–4
globalizations 60–1
Google 60–1
Memmi, Albert 50–1, 118–19, 137, 148, 166–7, 212–13, 218
methodologies, *see under* research
migration (environmental), *see* environmental migration *under* environmental problems

Nature 139–40, *see also* world-Earth
citizenships (*see under* world-Earth)
dominance of Nature 181–2
evolution (*see under* evolving)
laws of Nature 23, 32, 99, 103–4, 152–3, 178–9, 186–7, 217, 240
limitations of knowing (*see* limitations of reflectivity *under* ecopedagogy)
neocoloniality, *see also* coloniality
citizenships 213–15 (*see also* coloniality *under* citizenships)
development (*see* coloniality *under* development)
globalizations 64–8, 165
myth (countering) (*see under* coloniality)
neoliberalism 73–5
neoliberalism
development 68–9, 85, 107–8, 155
economics 61–2, 69–75
Freire, Paulo 68, 70–1, 89, 107–8, 208
Global Citizenship Education (GCE) (*see under entry*)
globalizations 51, 61–2, 65, 68–75, 104–5, 208, 217–18, 223, 238
neocoloniality (*see entry*)
private *vs.* public sphere 72–3, 88, 153–4, 186, 209–10, 222–5

NGOs (Non-Governmental Organizations)
Bill and Melinda Gates Foundation (*see entry*)
globalizations 59–61
NIMBY 77, *see also* Singh, Neera M.
NIAMBY & NOPE 2
Nyerere, Julius 47–8

Organic Ecopedagogical Modeling (OEM) 39–42, 49, 51, 54–5
Orientalism, *see* Said, Edward

peace education 126
conflict 29–30, 115–16, 119–20
pedagogies on the environment, *see under* environmental pedagogies
permaculture, *see* industrial Farming *under* environmental problems
Peters, Michael A. 219–20, 232–3
Planetary Citizenship Education (PCE) 23–4, 76–7, 201–2
globalizations 62–3
Plato's cave 232–3
politics, *see under* Freire, Paulo
positionality 145–52
fishbowl question (*see under* research)
oppressor-oppressed positioning 149
privilege problematizing 150–1 (*see also* limitations of reflectivity *under* ecopedagogy)
research 146–7
post-coloniality, *see* coloniality
post-truthism 219–20, 229–42
development 233
epistemologies (*see* entry)
globalizations 231
sciences and technologies (*see under entry*; media)
subjectivity 55–6
theorizing 164–5
Trump, Donald J. (*see entry*)
praxis 45, 110, 160–1
Freire, Paulo 123–31, 139–40
problematizing, *see* problem-posing *under* ecopedagogy *and under* Freire, Paulo

race, *see* critical race theories *under* theorizing
radicalness, *see* radical essence *under* ecopedagogy
Reaganism and Thatcherism 56–7
research 188–99
bottom-up 193–5
as curiosity (a la Paulo Freire) 55, 193–4
ecopedagogical spaces as research spaces 196–9
fishbowl question 47, 54–5, 145–6
for this book 11, 110–11, 136, 143–4, 155–7, 208–9
methodologies 177–9, 198–9
positionality (*see under entry*; fishbowl question *under this entry*)
subjectivity and objectivity 47–52
Rhodes, Cecil, *see* South Africa *under* environmental problems

Said, Edward 25, 50–1, 167–8, 212–13
Santos, Boaventura de Sousa, *see* de Sousa Santos, Boaventura
sciences and technologies 52–6, 158–9, 177–9
absences/emergences of disciplines (*see under* ecopedagogy)
dominance 52–4, 171–2
Freire, Paulo 178–9, 199
hard sciences problematizing 188–9
indigenous/local *vs*. 'modern' 183–4 (*see also* epistemologies)
post-truthism 230–1
Singh, Neera M. 2–3, 77–9, 184, 224–5
Smith, Graham Hingangaroa 183–4
Smith, Linda 183–4
social media, *see* media
social-environmental movements 31, 57–8, 69, 79, 155, 166, 182, 224
solidarity, *see under* Boaventura; citizenships; de Sousa Santos
Sontag, Susan 232–3
sustainability
baselines 84–8, 149
Education for Sustainable Development (ESD) (*see entry*)
epistemologies 180–1
literacy (*see* sustainability (introduction) *under* ecopedagogical literacies)
Sustainable Devleopmen Goals (SDG) (*see entry*)
world-Earth (*see under entry*)
Sustainable Development Goals (SDG) 13, 229, 242–6

technologies, *see* sciences and technologies
Thatcherism, *see* Reaganism and Thatcherism
theorizing 6, 10, 25, 144–5
  coloniality (*see entry*)
  critical media culture theories (*see under* media)
  critical race theories 3, 31–2, 66, 144–8, 182, 213–14
  feminism (*see entry*)
  lenses as a metaphor 161
  post-truthism (*see entry*)
  praxis (*see entry*)
  quilting theories (*see under* feminism)
  theoretical frameworks construction 31–2, 161–5
  trans-theorizing 66–7, 160–72
Tierney, Robert J. 54, 63, 65, 148, 151, 164, 167–71, 177–8, 184, 192–3, 215, 237–8
Torres, Carlos Alberto 33–6, 82–3, 96–8, 134, 154, 185–6, 214–15, 219–20
traditional ecological knowledge (TEK), *see* indigenous/local *vs.* "modern" *under* sciences and technologies; epistemologies
transdisciplinary 10, 46–7, 156–60, 188–99
  absences/emergences of disciplines (*see under* ecopedagogy)
  curricula 156–7
  false separation of disciplines 159
transformation 120–1, 123–31, *see also* Freire, Paulo
  vs. development or evolution (*see* evolving (*vs.* development/ transformation) *under* development)
transnational corporations (TNCs) 237
  globalizations 59–61, 203
trans-theorizing, *see under* theorizing
Trump, Donald J. 72, 233–8

utopia and education, *see under* Freire, Paulo

violence, *see* peace education

Westernization 67–8, 170–1
  Western individualism 48–9, 221–2
Western modern Science (WMS), *see* epistemologies; indigenous/local *vs.* "modern" *under* sciences and technologies
White, Hayden 230, 241–2
World Bank 59–61
world-Earth 6–7, 17–22, 32–42, 44–89, *see also* Nature
  citizenships 201–2
  distancing (*see under entry*)
  dominance of Nature (*see under* Nature)
  evolving (*vs.* development/ transformation) (*see under* development)
  Freire, Paulo 97–8, 124
  laws of Nature (*see under* Nature)
  limitations of knowing Nature (*see* limitations of reflectivity *under* ecopedagogy)
  Organic Ecopedagogical Modeling (OEM) (*see entry*)
  sustainability 23, 79–80, 224–5

zero-sum goals, *see under* development

CPSIA information can be obtained
at www.ICGtesting.com
Printed in the USA
LVHW051752221020
669549LV00006B/94

9 781350 083790